Eduardo Pagel Floriano

Wildlife conservation and management

Eduardo Pagel Floriano

Wildlife conservation and management

threats and protection measures

ScienciaScripts

Imprint

Any brand names and product names mentioned in this book are subject to trademark, brand or patent protection and are trademarks or registered trademarks of their respective holders. The use of brand names, product names, common names, trade names, product descriptions etc. even without a particular marking in this work is in no way to be construed to mean that such names may be regarded as unrestricted in respect of trademark and brand protection legislation and could thus be used by anyone.

Cover image: www.ingimage.com

This book is a translation from the original published under ISBN 978-620-6-76036-8.

Publisher:
Sciencia Scripts
is a trademark of
Dodo Books Indian Ocean Ltd. and OmniScriptum S.R.L publishing group

120 High Road, East Finchley, London, N2 9ED, United Kingdom
Str. Armeneasca 28/1, office 1, Chisinau MD-2012, Republic of Moldova, Europe
Printed at: see last page
ISBN: 978-620-7-67715-3

F635c Floriano, Eduardo Pagel
      Wildlife conservation and management / Eduardo Pagel Floriano. - Rio
Largo, 2024.
      206p.; ill.

      Includes bibliography and annexes
      ISBN

      Endangered species, valuation of nature, endangered animals, red list.
I. Title.

                                              CDU: 502.74

PRESENTATION

Many people born until the 1960s were hunters and fishermen, which was considered natural. Humans were gatherers and hunters at the dawn of civilization and hunting only ceased to be acceptable from the middle of the 20th century onwards. Hunters and fishermen acquired practical knowledge about wildlife, which made them more efficient and allowed those who were more aware of the natural limitations of animal populations to self-regulate their hunting and fishing efforts, sometimes based on data from research carried out by scientific institutions. In Rio Grande do Sul, for example, game fauna management was carried out until the beginning of the 21st century; the state was divided into four hunting regions; each year amateur hunting was allowed in one of them, but only after a fauna inventory had been carried out by the Zoobotanical Foundation of Rio Grande do Sul, which made it possible to determine the hunting effort that would be allowed. There were many hunting and fishing clubs and their members, who contributed an annual fee used for inspection, subsidizing other actions for the planning of game hunting, also participated in hunting inspection, reinforcing the contingent of state inspectors. In 2005, hunting was banned, but it is still illegal and uncontrolled, in a situation that is considered worse than the way hunting was managed.

The study of the conservation and management of fauna in the Forest Engineering course in its early days in Brazil provided some knowledge on the subject, mainly about the management of game species. In the 1980s, I took part in a team to draw up a fauna management plan for a property in Santa Catarina, which gave us a better understanding of the subject through visits to institutions and interviews with professionals who were already working in the area, mainly from the Zoobotanical Foundation of Rio Grande do Sul, which was carrying out the fauna inventory in that state and planning the management of game. Later, in the mid-1990s, with the installation and licensing of a conservation breeding facility in Santa Rosa, RS, we began to better understand the problems of breeding wild animals in captivity.

In addition to the scientific research carried out by universities and research institutions around the world, naturalists such as Jacques-Yves

Cousteau and David Frederick Attenborough have unraveled many of the mysteries of wildlife, as have the BBC, National Geographic and WWF with their various studies and documentaries on the natural environment. Today, thousands of amateur and professional videos, including scientific ones, shed light on wildlife and the problems caused by civilization. There are also a good number of books on environmental conservation and preservation and the impacts of anthropization. A milestone is Jean Dorst's 1971 work "Before Nature Dies", which he wrote:

> "Modern man is squandering non-renewable resources, natural fuels and minerals, without worrying about the future, thus running the risk of bringing about the ruin of today's civilization. Renewable resources, those we extract from the living world, are being squandered with bewildering prodigality, which is all the more serious because it could lead to the extermination of the human species itself: man can do without everything except food (Dorst, 1971)."

The anthropization of the environment and its consequences are already well known, and we need to use this knowledge. The Environmental Management course I completed in 2003 shed light on many ways of reducing and controlling the impacts we cause. Then, in 2021, there was a shortage of specialists to teach Wildlife Management on the Forestry Engineering course at the Federal University of Alagoas, which I took on with some fear of not having the most up-to-date knowledge required. I had to search for literature, laws, videos and articles to be able to teach the subject. From this study came this text, which I hope will be useful to our students.

Maceió, January 21, 2024.

Eduardo Pagel Floriano

1 INTRODUCTION

In the 1950s and 60s, awareness of the environmental problems resulting from human activity increased significantly. Many people became concerned about the environmental impacts of industrialization, urbanization and the exploitation of natural resources. Disasters drew worldwide attention to environmental dangers, such as the mercury pollution at Minamata in Japan and the Exxon Valdez oil spill in the United States. The books "Silent Spring" by Rachel Carson, published in 1962, raised the damaging effects of pesticides, and "The Limits to Growth" by Donella H. Meadows, Dennis L. Meadows, Jørgen Randers and William W. Behrens III, commissioned by the Club of Rome, showed the consequences of the exponential growth of the world's population, warning of the finite limits of natural resources and influencing public opinion. Environmental activism grew and several countries began to adopt protective measures, creating a context for international collaboration and discussion on a global scale.

The new awareness of the environment led to the United Nations Conference on the Human Environment, held in Stockholm in 1972, which raised the importance of preserving and conserving the environment, resulting in the declaration of 26 principles for good environmental management, including the Stockholm Declaration and Plan of Action for the Human Environment and various resolutions. From the discussions at the conference, the United Nations Environment Program (UNEP) was created.

Interactions between geological activities and living beings shape the climate and soils, resulting in the evolution of the natural environment. Prehistoric hunters and gatherers had little influence on the natural environment. But since the domestication of plants and animals, our influence has increased and has grown exponentially with the invention of tools and machines that have facilitated the production of food, buildings and transportation routes that have allowed the human population to increase. We are destroying the natural environment and exterminating species that may be crucial to our survival.

Living beings are classified according to kingdom, phylum, class, order, family, genus and species. The animal kingdom is made up of around 36 different phyla (SIDDIKA et al, 2016) which are subdivided into many classes. The best-known phyla are represented in Figura 1. Brazil has around 6.67% of the total described species, close to 100,000 of the 1.5 million known species, and a large number of species have not yet been described.

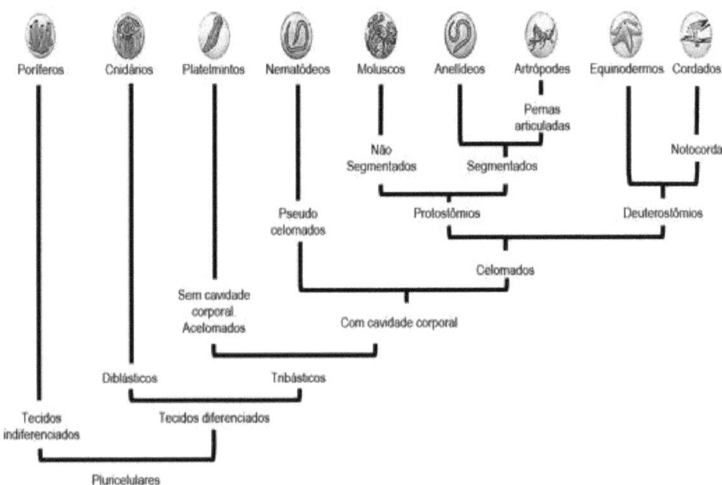

Figura 1 - Phylogenetic classification of the kingdom Animalia. Source: Siddika et al (2026).

Predatory animals at the top of the trophic chain are very susceptible to environmental changes made by humans. The overexploitation of wild species as food, which are part of the wild prey group, reduces their populations and puts them and their predators at risk of extinction. As a result of the reduction in predators and prey, animal populations and natural communities in general are out of control, with some reproducing excessively and others being reduced to critical levels. Therefore, studying anthropogenic changes and their consequences on the ecological relationships between wild species is crucial to recovering and conserving them. The study of animal ecology allows us to understand intra- and interspecific interactions, adaptation, geographical distribution, population density, the evolution and coevolution of their populations and the processes of natural selection of species.

The United Nations Conference on Development and the Human Environment, which took place in Stockholm from June 5 to 16, 1972, with the participation of 113 countries, can be considered the main milestone in global environmental awareness. At the conference, there was international integration for nature conservation, although many countries had already issued environmental protection standards at the time. New emphasis on environmental protection was given by the United Nations Conference on Environment and Development, held in Rio de Janeiro from June 3 to 14, 1992, addressing the consequences of exponential population growth, air pollution, intense exploitation of natural resources and the serious consequences for the future of humanity.

Wilderness is the result of geological evolution in communion with living beings and the climate. Natural changes are part of the evolution of environmental conditions, but they maintain balances over relatively long periods of time that have only been broken by natural disasters and now by the actions of human beings. Living beings, soil, geology and climate have interacted for billions of years to create a reasonably stable environment. Living species have their own specific ecosystem functions, such as breaking down rocks, fixing nitrogen that is useful for higher plants, feeding other species, pollination, climate regulation, protecting soil and water sources, dispersing seeds, decomposing organic waste, providing shelter for other species, regulating populations, etc.

Specifically for humans, wild animals are important in relation to:

- Balancing natural ecosystems;
- Tourist exploitation;
- Environmental education;
- Food: free-living fish, protein supply for traditional populations.

Animals bred or exploited by humans were once wild and some still are, having been adopted by human civilization, which benefits from them in different ways:

- Companionship and security - pets and guide dogs, sniffer dogs, guard dogs;
- Food - honey, meat, collagen gelatine, etc;
- Clothing - wool, leather, fur, feathers;
- Medicines - heparin, captopril, tablet and lozenge coatings, syrups, etc;
- Cosmetics and cleaning - animal fat soap, collagen creams, elastin moisturizers, hair conditioners and shampoos with keratin, biotin and retinol; preservatives based on animal urea, carmine dyes, perfumes;
- Energy generation - methane from pig waste;
- Fertilizers - composting;
- Food for other farm animals - Animal feed;
- Other products of animal origin: stearic acid, caprylic acid, casein, animal fat (lard), animal hair and skin, musk, ambergris, castoreum, civet, hyaluronic acid, allantoin, lanolin, fish glue, etc (AUR, 2021).

We are at the top of all food chains, we use animals and animal products for different benefits and any extinct species has an impact on human civilization. Protecting them, in addition to the ethical aspect of being the only species that has the power to consciously and significantly modify the environment, is essential for our survival and well-being.

2 LEGISLATION

There was little concern about wild fauna in Brazil until 1924, so much so that Law No. 3,071, of January 1, 1916, considered fauna to be private property, belonging to the landowner, and dealt with it in the sections "On Hunting" and "On Fishing", as transcribed below:

"THE HUNT

Art. 594. Subject to administrative hunting regulations, hunting may be carried out on public land or on private land, with the license of its owner.

Art. 595. The animal seized by the hunter belongs to the hunter. If the hunter goes after the animal and has wounded it, it shall belong to him, even though someone else has seized it.

Art. 596: Domestic animals that run away from their owners while the latter are looking for them are not considered hunting animals.

Art. 597: If wounded game is brought onto fenced, walled, gated or cultivated land, the owner of the land, unwilling to allow the hunter to enter, must hand it over or expel it.

Art. 598: Anyone who enters someone else's land without the owner's license to hunt shall forfeit to the owner the game he catches and shall be liable for any damage he causes.

FISHING

Art. 599. Subject to administrative regulations, it is lawful to fish in public waters, or in private waters, with the consent of their owner.

Art. 600. The fish that he catches belongs to the fisherman, as does the fish that the harpoon or barb chases, even if someone else catches it.

Art. 601. Anyone who fishes in someone else's waters without the owner's permission shall forfeit the fish he catches and shall be liable to him for any damage he causes.

Art. 602. On private waters that cross land owned by many people, each of the riparians has the right to fish on their side until noon."

The "Appropriation" section of the same law regulates the appropriation of unowned animals, as follows:

"APPROPRIATION

Art. 592: Anyone who takes possession of something that has been abandoned or has not yet been appropriated immediately acquires ownership of it, and this occupation is not defended by law.

Sole paragraph. Movable property becomes ownerless when its owner abandons it with the intention of renouncing it.

Art. 593: These are ownerless things subject to appropriation:

I - Wild animals, while left to their natural freedom.

II - The tame and domesticated who are not marked, if they have lost the habit of returning to the place where they usually gather, except in the case of art. 596.

III - Swarms of bees previously appropriated, if the owner of the hive to which they belonged does not claim them immediately.

IV - Stones, shells and other mineral, vegetable or animal substances washed up on beaches by the sea, if they show no sign of previous ownership."

The first Brazilian legal act on animal protection was Decree 16.590 of 1924, which regulated practices in public amusement houses, banning animal races and fights, among other amusements at the expense of animal abuse (SENA, 2021).

In 1934, there was a change in animal protection with Decree No. 23.672, of January 22, 1934, which placed all animals in the country under the guardianship of the state, also establishing a ban on mistreatment of animals, including by their owner or possessor. It also sought to regulate hunting and fishing activities, establishing a ban on professional hunting without state registration and hunting without the owner's permission in areas of public or private domain. Decree 24.645 was also enacted in 1934, establishing measures to protect animals and defining what mistreatment meant in 31 sections.

In 1967, the Fauna Protection Law was enacted, Law No. 5,197 of January 3, 1967, which states in Article 1 that: "Animals of any species, at any stage of their development and living naturally outside captivity, constituting wild fauna, as well as their nests, shelters and natural breeding grounds are the property of the State, and their use, persecution, destruction, hunting or gathering is prohibited."

Law No. 5.197/67 is still in force, but has been modified by the following legal acts:

- Law 7.584, of 06/01/1987: adds and renumbers items I and II of §2 of Art. 33;
- Law 7.653, of 12/02/1988: amends Arts. 27, 33 and 34;
- MPV 10, of 21/10/1988: revokes § 4º of Art. 27;
- Law 7.679, of 23/11/1988: repeals § 4 of Art. 27;
- Law 9.111, of 10/10/1995: adds § 3 to Art. 3;
- Law 9.985, of 18/07/2000: repeals Art. 5;

Other important acts related to the Fauna Protection Law are the following:

- Law 7.173, of 14/12/1983: application of Art. 1;
- Decree 97.633, of 10/04/1989: provides for the National Fauna Protection Council - CNPF;
- Decree 97.946, of July 11, 1989: basic structure of IBAMA;
- Decree 1.218, of 15/08/1994: delegation of powers;
- Ordinance/MMA No. 241 - Official Gazette of August 31, 1994, p. 13157: CNPF internal regulations;
- Portaria/MMA Nº 23 - D.O. de 26/02/1999, p. 131: approves the Internal Regulations of the National Fauna Protection Council - CNPF, created by Art. 36 of Law 5.197, of 1967;
- Decree 3.179, of 21/09/1999: specifies the sanctions applicable to conduct and activities harmful to the environment.

Within the scope of regulatory bodies, the following standards can be cited:

- Ibama Ordinance No. 93, of July 7, 1998 - deals with the Import and Export of Wild Fauna. Defines domestic, exotic and wild fauna and regulates the transportation, trade and possession of pets.
- Ibama Ordinance No. 117, of October 15, 1997 - Standardizes the sale of live animals, slaughtered animals, parts and products of

the Brazilian wild fauna from breeding facilities for economic and industrial purposes and zoos registered with IBAMA.

- Ibama Normative Instruction No. 31, of December 31, 2002 - Provides for the temporary suspension of requests for commercial breeding facilities for reptiles, amphibians and invertebrates for the purpose of producing pets for sale on the domestic market, and makes other provisions.
- Ibama Normative Instruction No. 7, of April 30, 2015 - Establishes and regulates the categories of use and management of wild fauna in captivity, and defines, within the scope of IBAMA, the authorization procedures for the established categories.
- Conama Resolution No. 394, of November 6, 2007 - Establishes the criteria for determining wild species to be bred and marketed as pets.

3 VALUING NATURE

Valuing nature is very subjective. It is difficult to define the value of a species, its environmental services or a territorial unit, because few environmental goods are commercialized. A few goods, such as carbon stocks and water, have commercial values, but their environmental value may be different from their commercial value, depending on the environmental services they provide. Valuing ecosystems and nature conservation units is no easy task either. Estimating the economic value of goods and services provided by nature is still an evolving science.

The International Union for Conservation of Nature (IUCN) began a study into the valuation of nature in the 1990s, setting itself the mission of looking at the issue of economic value, i.e. the types of economic value generated by conservation activity that may not be captured on the market, and failed in its attempt to provide a system to help conserve natural resources by following this path. He then began to question why "biodiversity disappears and how its economic value can be captured by various institutional mechanisms". As a result, in 1994 he published a new study entitled "THE ECONOMIC VALUE OF BIODIVERSITY", authored by David Pearce and Dominic Moran in association with the Biodiversity Program of IUCN - The World Conservation Union:

> "- economic forces drive much of the extinction of the world's biological resources and biological diversity; yet
>
> - biodiversity has economic value. If the world's economies are rationally organized, this suggests that biodiversity should have less economic value than the economic activities that give rise to its loss;
>
> - However, we know that many biological resources have significant economic value. We also know that many of the destructive activities themselves have a very low economic value; therefore
>
> - something is wrong with the way real economic decisions are made - for some reason, they fail to "capture" the economic values that can be identified;

- These "economic failures" are at the heart of any explanation for the loss of biological diversity. If we can tackle them, there is a chance of reducing biodiversity loss (PEARSE and MORAN, 1994)."

The attribution of economic value to biodiversity is based on the argument that it is a practical method for its conservation. However, great care must be taken, as the economic values generated by conservation activities may not be captured on the market, generating distortions that could lead to the destruction of biological resources (PEARSE and MORAN, 1994) , which is the opposite of what is desired.

When valuing biodiversity, its intrinsic value must be taken into account, as well as nature's contributions to people's quality of life, in addition to its economic value, which were taken into account by Schröter et al (2020). The intrinsic value of biodiversity is that which living beings have in themselves, unrelated to human use and generally not measurable. The extrinsic value is that which is related to current or potential future human use, schematized in Figura 2.

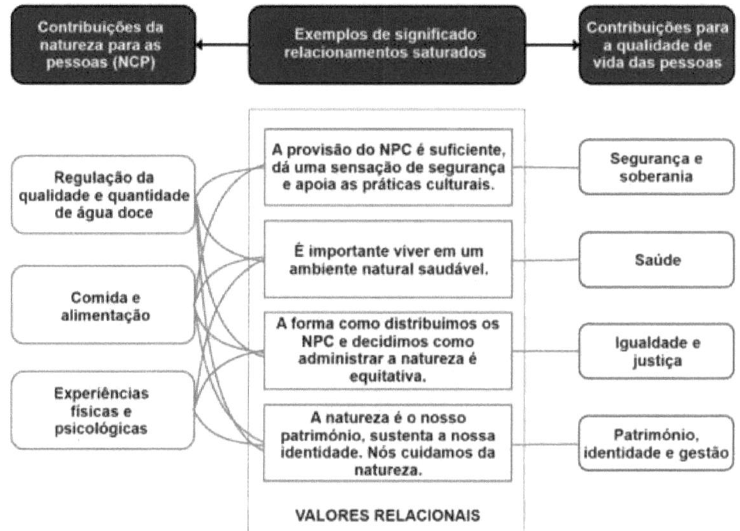

Figura 2 - Nature's contribution to people's quality of life (SCHRÖTER et al, 2020).

Biodiversity evolves through interaction with ecological processes to form different ecosystems. The main interactions between biodiversity and ecological processes in the biosphere are shown below in Figura 3.

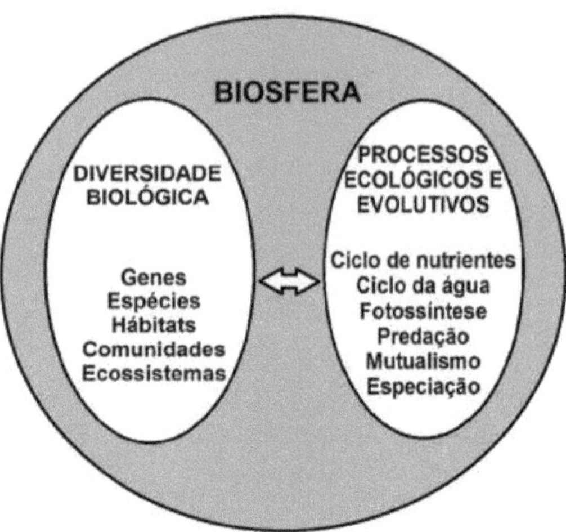

Figura 3 - Relationships between biological diversity and ecological processes. Source: Reid and Miller (1989).

Some of the approaches and concepts used to measure biodiversity and its components are mentioned in the following sections.

3.1 Genetic diversity

Biodiversity includes species diversity, genetic diversity and ecosystem diversity (MELO, 2021).

Biological diversity is described as the richness and variability of genes, species and ecosystems, which correspond to the three fundamental levels of biological organization. Genetic diversity is all the genes of all living beings on Earth and all the genetic information they contain (PEARSE and MORAN, 1994).

Every living being is made up of a number of genes that are repeated in different arrangements, but some genes can have exclusive mutations, which makes them unique; in other words, the loss of an individual from a single species can mean the loss of important genes for good, as the mutations that imprinted those characteristics may never occur again. In some cases, the significant loss of individuals from a population can lead to genetic narrowing and inbreeding, making the survival of the population unviable because it doesn't have enough genetic variability to keep it healthy.

3.2 Diversity of species

A biological species is a group of similar organisms capable of reproducing and generating fertile offspring. Individuals of different species sometimes produce hybrid offspring, but these hybrids are unable to reproduce among themselves, or with individuals of the original species. This is called reproductive isolation and prevents gene flow between different species. Gene flow only occurs between individuals of the same species, but there are exceptions.

New species can emerge as a result of geographical isolation in different environmental conditions and return to coexist as distinct species unable to reproduce with each other. Gradients and changes in species richness are correlated with existing environmental variations, such as precipitation, nutrient availability, salinity, among other variations in climate and available energy. The number of existing species is unknown. Around 1.5 million species have been described and it is estimated that there are more than 5 million, which could be as many as 50 million (PEARSE and MORAN, 1994).

3.3 Ecosystem diversity

A habitat is formed by a set of environmental conditions favorable to the life and development of a particular animal or plant species.

The different ecological conditions of each location make it possible to form different habitats where living beings live that are able to survive in these environments. Different habitats allow for the development of different biotic communities that make up ecosystems.

Ecosystem diversity, therefore, relates to the variety of habitats, biotic communities and ecological processes that exist within ecosystems and can be described at different levels as follows (PEARSE and MORAN, 1994):

> ➢ Functional diversity - relative abundance of functionally different types of organisms;
>
> ➢ Community diversity - the number, size and spatial distribution of communities, referred to as patches.
>
> ➢ Landscape diversity - variation in the scale of landscape patches.

Reid and Miller (1989) cite six factors that influence ecosystem dynamics and link environmental change, biodiversity and ecosystem processes:

1. "The mix of species that make up communities and ecosystems changes continuously, even though they may appear stable.

2. Species diversity increases as environmental heterogeneity - or the fragmentation of a habitat - increases, but although species richness can sometimes be increased by increasing the diversity of habitats within an ecosystem, this intervention can be a double-edged sword.

3. Habitat irregularity influences not only the composition of species in an ecosystem, but also the interactions between species.

4. Periodic disturbances play an important role in creating irregular environments that promote high species richness. Periodic disturbances maintain a series of habitat patches in various successional states.

5. Both the size and isolation of habitat patches can influence species richness, as can the extent of transition zones between habitats.

6. Certain species have disproportionate influences on the characteristics of an ecosystem."

Therefore, the concept of biodiversity is comprehensive and complex, involving various levels and scales.

3.4 Measuring biodiversity

Measuring biodiversity is not an exact science, it is subjective and controversial, depending on different levels and scales, as in the following four sections.

3.4.1 Measuring genetic diversity

The diversity of alleles present in a given population is responsible for differentiating the morphological and physiological characteristics of organisms, resulting in variations between the characteristics of each individual. Genetic differences can be measured in terms of phenotypic characteristics, allele frequencies or DNA sequences.

3.4.2 Phenetic diversity

The different morphological and physiological characteristics that organisms display are the basis of phenetic classification, disregarding phylogenetic relationships. Measures of easily measurable characteristics are generally used, which are divided into classes by their size. It is therefore based on measurements of phenotypes, which avoids examining the underlying allelic structure. Phenetic classification is useful both ecologically and practically.

3.4.3 Allelic diversity

Allele genes are those that are found in the same locus on homologous chromosomes and participate in determining the same character, whether or not they determine the same aspect.

Allelic diversity is measured at the individual or population level using allelic composition at individual loci, which can be obtained by protein electrophoresis.

A greater number of alleles leads to more equitable frequencies and the greater the number of polymorphic loci, the greater the allelic diversity.

3.4.4 DNA sequence variation

The polymerase chain reaction is used to obtain a small amount of sequenced DNA, with just a drop of blood as a sample, and to compare the DNA sequences of different living beings.

3.5 Measuring species diversity

Species diversity is a measure that involves the number of species and individuals of each species, i.e. the species richness and abundance of each species in a given region or ecosystem.

3.6 Measuring community diversity

The diversity of communities varies with the different habitats in the community, the relative abundance of species, the age structure of populations, the patterns of communities in the landscape, the trophic structure and the dynamics of patches (REID AND MILLER, 1989).

There is great complexity in measuring ecosystem diversity, resulting in different measures of diversity at the community level; for example: biogeographic kingdoms or provinces are defined on the basis of species distribution and ecoregions or ecozones on the basis of physical attributes such as soils and climate (Reid and Miller, 1989).

According to Reid and Miller (1989) these definitions can differ according to the scale:

> "For example, the world has been divided into biogeographical provinces or more refined classifications that may be more useful for policy-making. More policy-oriented measures include the

definition of 'hotspots', based on the number of endemic species and 'megadiversity' states."

3.7 Valuation of ecosystem units

A study to estimate the value of nature was carried out by the state of Paraná with the aim of physically expanding the protected areas and creating new PAs and increasing the expressiveness of the PAs already established, with a view to optimizing the conservation of biodiversity *in situ and* providing greater representativeness and stability of the PAs, which are indispensable in the search for the sustainability of life on the planet. Tossulino et al (2005) cite the criteria defined for the analysis of each PA according to a synoptic set of the main characteristics considered to define the management categories, as follows:

" The degrees of valuation defined to highlight the most representative characteristics of each area are described below:

Degree of ecosystem conservation (EC)

To value this characteristic, data on the size of the remnant of the original plant type and its conservation status were taken into account, as well as the presence of associated fauna and abiotic characteristics that are relevant to its maintenance. The values for the state of conservation of the ecosystems are as follows:

(i): ecosystems unchanged;

(ii): ecosystems that are little altered or recovering; and

(iii): de-characterized ecosystems.

Successional Stage (ES)

It is the analysis of the state of development of the remnant of the original plant typology, classifying it as original or in a state of regeneration.

The values are as follows:

(i): predominance of original plant types;

(ii): predominance of vegetation in a state of regeneration;

(iii): predominance of uncharacterized vegetation or vegetation in an initial state of regeneration.

Homogeneous reforestation (RF)

It indicates, quantitatively in relation to the total area, homogeneous reforestation with exotic and/or native species, as well as their representativeness for promoting sustainable forest management. The figures for the current state of reforestation are as follows:

(i): areas with 75 to 100% reforestation and only representative enough to carry out forest management studies;

(ii): areas with 50 to 75% reforestation and high representativeness for developing forest management studies;

(iii): area with 25 to 50% reforestation and with medium representativeness for developing forest management studies;

(iv): area with less than 25% reforestation and with low or no representativeness for forest management studies.

Endangered Species (EXT)

It indicates the presence or absence of endangered species on the list of threatened or endangered species of fauna and flora (Paraná, 1995a and Paraná, 1995b).

The figures for the presence of endangered species are as follows:

(i): proven occurrence of endangered species;

(ii): the possibility of the occurrence, even if possible, of endangered species;

(iii): no findings and no possibility of endangered species.

Scenic Beauty (BC)

It indicates the presence and representativeness of scenic attributes such as natural cavities, waterfalls, vegetation, rock formations and archaeological and/or historical sites. The values relating to the presence of scenic beauty in the PAs studied are:

(i): presence of rare, unique natural sites or sites of great scenic beauty;

(ii): presence of natural attributes of scenic beauty;

(iii): insignificant characteristics.

Public Visitation (VP)

It is an analysis of the representativeness of the Unit for developing environmental education and interpretation activities and/or outdoor recreation.

The values for this characteristic are as follows:

(i): representativeness for educational and recreational visits;

(ii): representativeness for educational visits;

(iii): representativeness for recreational visitation;

(iv): not at all representative of visitation.

With the characteristics of each area duly valued, a comparison was made with the main characteristics that a protected area must have in order to fall into a management category defined by Snuc (TUSSOLINO et al, 2005). "

Richard B. Norgaard, from the University of California, Berkeley, in the United States, is considered one of the founders and an ongoing leader in the field of ecological economics. Norgaard argues that valuing wild animals is a complex task, as it is necessary to consider various criteria, including the intrinsic and extrinsic values of the animals.

Intrinsic values - are those inherent to the animal itself, not dependent on external factors. They include:
- Aesthetic value: the beauty or uniqueness of the animal.
- Scientific value: the importance of animals for scientific research.
- Ecological value: the animal's role in the ecosystem.

Extrinsic values - are those that depend on external factors, such as human culture or the economy. They include:
- Cultural value: the importance of the animal to the culture of a people.
- Commercial value: the value of the animal on the wildlife market.
- Recreational value: the importance of animals for tourism.

Intrinsic criteria are considered more important for biodiversity conservation, while extrinsic criteria are more relevant for combating the illegal exploitation of wild animals.

Some examples of wild animal valuation criteria are listed below:
- A rare and endangered animal of great ecological importance would be considered to be of high intrinsic value.

> ➢ An animal that is considered sacred in a certain culture would have a high cultural value.
> ➢ An animal that is popular on the illegal wildlife market, it would have a high commercial value.
> ➢ An animal that is a tourist attraction would have high recreational value.

The economic values of biodiversity resources (Tabela 1) can be described as follows (UNEP/CBD/SBSTTA, 1996):
> ➢ Passive use value or non-use value is the value that individuals attribute to biological resources that they do not intend to use, but would feel the "loss" if they disappeared.
> ➢ Direct Value refers to benefits that can be observed being consumed, although their consumption may not generate a significant price that can be attributed to that benefit.
> ➢ Indirect Value refers to benefits that are not observed being consumed, but which are recognized as essential to the preservation and maintenance of ecosystems.
> ➢ Option value is the value attributed to the guarantee of future consumption of goods and services that produce direct and indirect value.
> ➢ The quasi-option value is the value of learning about the future benefits that would be prevented by the development or irreversible change of forests today. This takes into account the fact that current valuations are circumscribed by current knowledge of forest functions.
> ➢ Existence Value is the value attributed to an environmental asset regardless of its current or future "use". This incorporates the innate value of the forest *in situ*.

TABELA 1 - Examples of the economic value of biodiversity resources

VALUE OF USE			+	LIABILITIES OR
DIRECT VALUE	**INDIRECT VALUE**	**OPTION VALUE**		**NON-USE VALUE**
			QUASI-OPTION VALUE	**EXISTENCE VALUE**
Provision of basic resources: food, medicines, building materials, nutrients.	Provide support for economic activity and human well-being, e.g. watershed protection, waste storage and recycling, maintenance of genetic diversity and erosion control. Provides basic resources: e.g. oxygen, water, genetic resources.	Preservation of future direct and indirect use values		
Non-consumable uses: recreation.			Conservation of as yet unknown future uses	Forests as objects of intrinsic value, as a legacy, as a gift to others, as a responsibility (stewardship). Includes cultural, religious and heritage values.
Plant genetic resources.	Provide information benefits, such as scientific knowledge.			

Source: UNEP/CBD/SBSTTA (1996)

The set of valuation methods makes it possible to gauge the economic value of the natural resource, taking into account the ecosystem goods and services that the resource provides (PARRON et al, 2015).

4 ANIMAL ECOLOGY

4.1 Animal behavior

Animals' needs for locomotion, shelter and protection, nutrition, watering, reproduction, protection and feeding of offspring result in different ecological interactions (Tabela 2) that influence their behavior.

TABELA 2 - Ecological interactions between animals

Types of ecological relationships			Effects	
Intraspecific	Harmonics	Cologne	+	+
		Society	+	+
	Disharmonious	Cannibalism	+	-
		Competition	-	-
Interspecific	Harmonics	Commensalism	+	0
		Protocooperation	+	+
		Mutualism	+	+
		Tenancy	+	0
		Symbiosis	+	+
		Foresia	+	0
	Disharmonious	Amensalism	0	-
		Competition	-	-
		Slavery	+	-
		Parasitism	+	-
		Predation	+	-
		Herbivory	+	-

Ecological relationships between animals are influenced by the position of the species in the various trophic levels of food chains (Figura 4).

Species with the same feeding behavior make up a trophic level.

Plants are autotrophic beings, they produce their own food, feed on the physical environment and constitute the trophic level of **producers**, forming the base of food chains.

Heterotrophic beings do not produce their own food and are characterized as herbivorous consumers, omnivores, carnivores and decomposers. Herbivores are called **primary consumers** and feed on plants. Omnivores feed on plants and animals and are considered secondary consumers. Carnivores are **secondary consumers** and feed on other animals. **Decomposers** are mainly bacteria and fungi, which promote the decomposition of waste and dead individuals, returning nutrients to the environment and restarting the cycle.

Figura 4 - Food chains. Source: Batista (2011).

Food chains can form (Figura 5), in which there are one or more producers for one or more consumers (BATISTA, 2011).

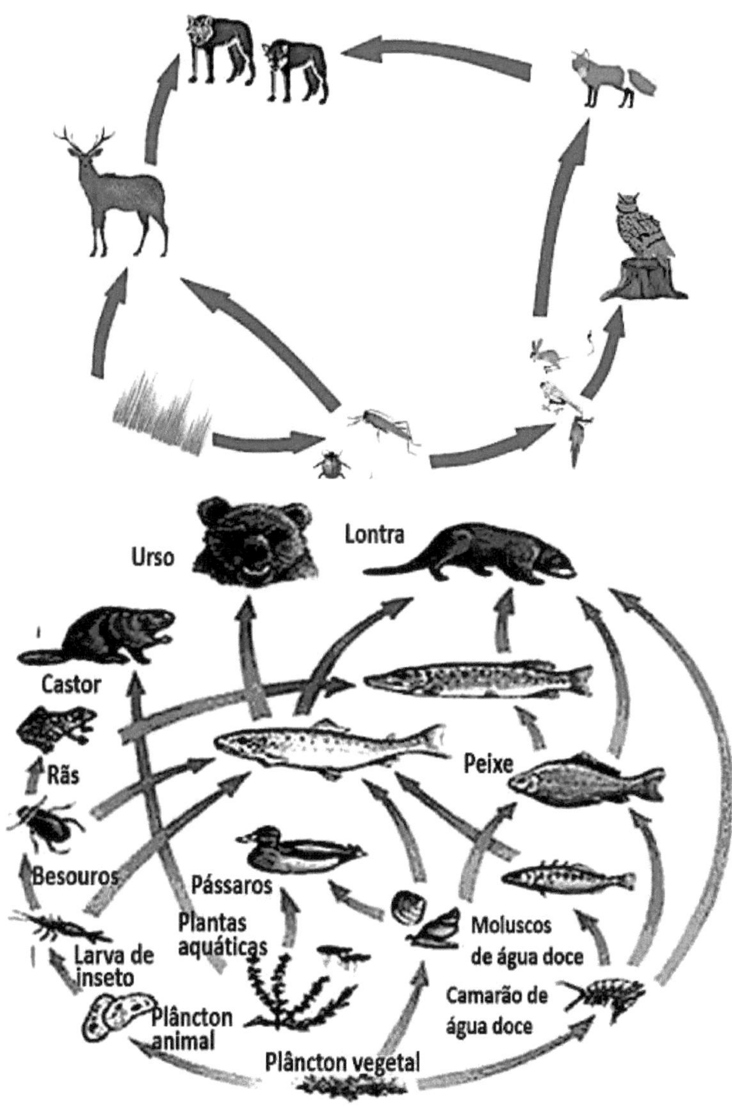

Figura 5 - Food webs. Source: Batista (2011).

4.2 Ecological interactions

A community is a collection of various populations of species that are structurally regulated through the ecological interactions that occur between them.

Ecological animal interactions can occur between individuals of the same species, known as intraspecific, and between individuals of different species, known as interspecific. Relationships between animals can be harmonious, when there is a benefit for at least one of the species involved in the relationship and none is harmed; or, disharmonious, when at least one of the species involved in the relationship is harmed.

4.2.1 Intraspecific harmonic relationships

4.2.1.1 Colonies

Colonies are associations between individuals of the same species that live together physically and anatomically, being interdependent, with benefits for all. Individuals cannot survive outside the colony (CECIERJ, 2016).

Colonies can be of two types:

- Homomorphic or isomorphic - all individuals have the same shape, no division of labor and all individuals perform all vital functions. Example: cnidarian corals formed by polyps with a calcium carbonate skeleton.
- Heteromorphic - the individuals are morphologically different and each has a distinct function that is vital to the colony. Example: the Portuguese caravel is a cnidarian of the *Physalia physalis* species made up of a colony of four different types of polyps (zooids), each with a different function, listed below: *pneumatophore* (air-filled vesicle for buoyancy); *domonoctozooids* (tentacles for catching prey); *gastrozooids* (colony stomachs); and, *gonozooids* (responsible for reproduction).

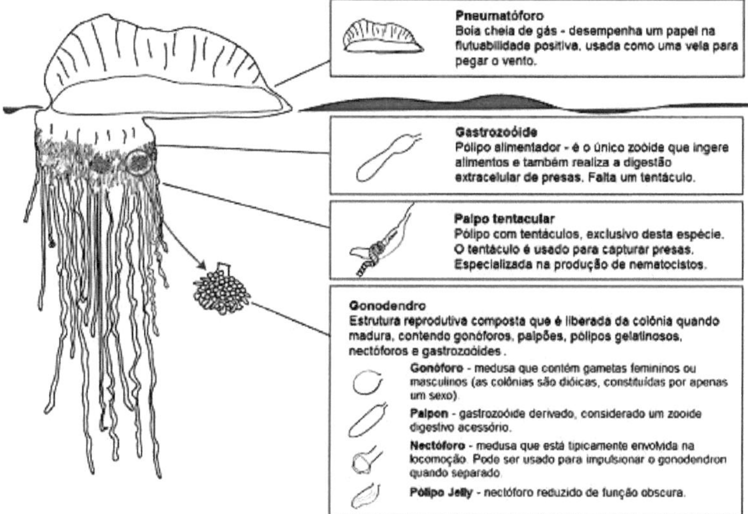

Figura 6 - Anatomy of a Portuguese caravel (*Physalia physalis* colony) with descriptions of the function of each zooid (polyp). Source: Munro et al (2019).

4.2.1.2 Society

Societies are associations of individuals of the same species, who live in cooperation without being anatomically linked, who come together due to reciprocal stimuli, with gregarious behavior and a strong tendency to live together (CECIERJ, 2016). They are usually made up of castes of individuals with different functions or age groups who share tasks and cooperate with each other, with benefits for all. Examples: termites, ants, bees (Figura 7), wild pigs, wolf packs. Social behavior involves complex communication between individuals and represents an evolutionary advantage, making it easier to obtain food, care for offspring and protect all members of the group.

Figura 7 - A typical example of a society is that formed by bees.

4.2.2 Harmonic interspecific interactions

4.2.2.1 Mutualism

It is the obligatory association between individuals of different species in which both benefit (SILVA and SANTOS, 2022). The survival of individuals depends on association, since it is impossible to live individually, but there is no need for the physical integration of individuals who generally provide complementary resources or services to each other. A typical example is the pollination of plants by bees; bees need plant nectar to produce honey and plants need bees for pollination.

4.2.2.2 Protocooperation

Protocooperation, also known as cooperation, is a bilateral association between individuals of different species in which there are benefits for both, although the two species can survive separately (SILVA and SANTOS, 2022). Examples: a heron feeds on cattle ectoparasites, both benefit, but don't need to live together.

Protocooperation and mutualism differ only in that in protocooperation the association is not obligatory, while in mutualism there is an obligation for the individuals of the two species to live together in order to survive.

4.2.2.3 Commensalism

It is the association between individuals of different species in which one of them takes advantage of the other's food remains without harming it (CECIERJ, 2016). The animal that takes advantage of the food remains is called a commensal. An example of commensalism that is often cited is that between the rémora and the shark. The rémora or lumpfish is a bony fish that has its dorsal fin transformed into a suction cup, with which it attaches itself to the shark's body. In addition to being transported by the shark, the rémora also uses the remains of its food. The shark is not harmed because the weight of the rémora is insignificant. The food ingested by the rémora corresponds to that discarded by the shark.

4.2.2.4 Tenancy

It occurs between individuals of different species in which one of them uses the other for shelter or support, without harming it (SANTOS, 2023). There is no food provided by one or the other in this association. One of the individuals is called the tenant and the other is the host. A classic example is the association between the needlefish and the sea cucumber; the needlefish seeks protection by sheltering inside the echinoderm's body when it is in danger. The hermit crab, which uses snail shells to shelter and protect its soft parts, is also considered a tenant. Another case is corals, which provide shelter for eels.

4.2.2.5 Metabiosis

Metabiosis "is the ecological relationship in which the existence of one organism creates the necessary conditions for the survival of another" (TUNES, 2020), even if it doesn't provide shelter as in the case of tenancy. An example of this is whales that die in the open sea and submerge to the bottom, where they create a complex ecosystem, supporting various species for decades.

4.2.2.6 Symbiosis

Symbiosis differs from mutualism in that the individuals interact physically, whereas in mutualism the individuals of the two species are not physically connected. Most of the time the benefit is mutual, but it can happen that one of them is harmed. Symbiosis in the animal kingdom encompasses a set of interactive biological relationships that are beneficial to one of the species involved. For interactions to be considered symbiotic, they must persist for a long time. In a symbiotic relationship, organisms actively act together for mutual benefit (SILVA and SANTOS, 2022).

Generally speaking, there are several classes of symbiosis, divided into two large groups:

- Endosymbiosis - This occurs when one of the living beings taking part in the relationship lives inside the other organism. They usually settle inside cells, in their digestive system. Example: Wood termite and protozoa - the protozoa digest the cellulose ingested by the termite, transforming it into glucose with the action of enzymes and both use glucose as a source of energy.
- Ectosymbiosis - Occurs when symbiotic beings live together, but without being inside their organisms. Generally, one species lives under the body of the other. Example: clownfish and sea anemone - the fish is not affected by the anemone's poison, which protects it from predators, and the fish provides the anemone with nutrients through its feces.

4.2.2.7 Foresia

It is the association between individuals of different species in which one uses the other for transportation, without harming it (CASSINI, 2005). It can be understood as a particular case of commensalism. In foresy, the relationship is not permanent; an individual simply hitches a temporary ride with the host (Figura 8).

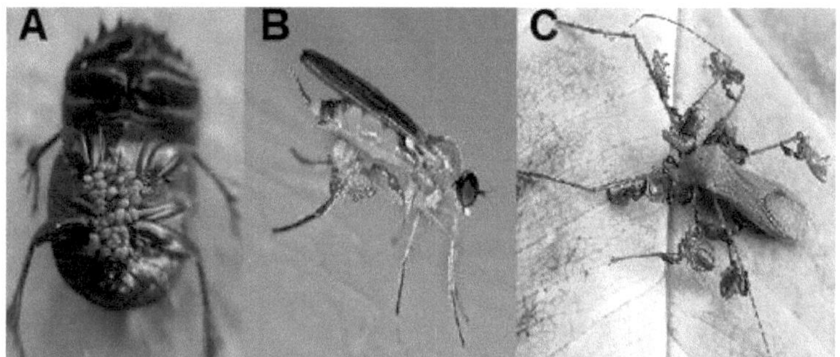

Figura 8 - Examples of foresy: A) ticks hitchhiking on beetles; pseudoscorpions hitchhiking on - B) wasps; C) bedbugs. Source: Sanches (2020).

4.2.3 Inharmonious interspecific relationships

4.2.3.1 Amensalism or antibiosis

It is the inharmonious interaction where one species produces and releases substances that hinder the growth or reproduction of others and can even kill them (CASSINI, 2005).

Figura 9 - An example of amensalism: herds of moving animals compete for food and trample on small animals and plants, reducing their populations.

In amensalism, one population is adversely affected by another, but the latter is neither benefited nor harmed (CECIERJ, 2016). Example: herds of large herbivorous animals grazing or moving around, which can trample and kill a large number of insects and affect their population, but the herbivores are not affected. Example of amensalism: red tide in which toxins are released by dinoflagellates, harming other species in the marine environment.

4.2.3.2 Predation

It is a disharmonious interaction in which one animal (predator) kills another animal of a different species (prey) for food (CECIERJ, 2016). Predators can be considered beneficial to the prey population, controlling their populations, eliminating sick and weak individuals, favoring natural selection.

Figura 10 - Example of predation: Jaguar preying on capybara.

The evolution of prey and predators is due to the species' need to survive. Prey have developed strategies of abundant reproduction to maintain large healthy populations, mimicry to hide, speed to escape, acute hearing and smell to identify the presence of predators, among others; while predators have developed teeth and claws to trap prey, colors for camouflage, keen eyesight, etc.

Camouflage can be a strategy of both prey and predators, who develop shapes (homotypy) or colors (homochromy) similar to the environment, which allow them to go unnoticed.

Mimicry can also be adopted by prey and predators and disguises them, making it difficult for opponents to detect them.

4.2.3.3 Parasitism

It is the inharmonious association between individuals of different species in which one lives at the expense of the other, harming it (CECIERJ, 2016). The individual that harms is called a parasite or biont. The harmed individual is called the host or biospecies.

Depending on the intensity of the parasitism, the parasite can cause the death of the host and, when this is obligatory, it is called parasitoidism, which is a special type of parasitism. Parasitoids almost always lead to the death of the host. This characteristic has been used for the biological control of pests.

Parasites can be classified in relation to the number of hosts as monogenetic when they use a single host to complete their life cycle, such as roundworms, or digenetic when they pass through more than one host to complete their life cycle, such as *Schistosoma and Trypanosoma*.

They are also classified according to their location in the host, being considered ectoparasites when they attach themselves externally to the host (e.g. leeches, ticks), or endoparasites when they parasitize the host internally, such as tapeworms and roundworms.

The host is simply an ephemeral resource that the parasite uses to maximize its fitness, either by increasing the probability of completing its life cycle or by increasing its ability to produce offspring. The way in which the parasite exploits the host as a resource will depend on which of the possible strategies maximizes its fitness, without any concern for the host's well-being (POULIN, 2007). The forms of host exploitation used by parasites can be summarized as: parasitic castration, directly transmitted parasitism, trophically transmitted parasitism, vector-transmitted parasitism, parasitoidism and micro predation.

4.2.3.4 Slavery or Sinophilia

It is the inharmonious interaction in which one species captures and makes use of the work, activities and even food of another species (CASSINI, 2005). Examples include certain Amazon ants and mat ants. One example is the relationship between ants and aphids. Aphids are parasites of certain plants. They feed on the sap they take from the vessels of plants such as rose bushes, orchids, etc. This sap is rich in sugars and poor in amino acids. Because they absorb a lot of sugar, aphids eliminate excess sugar through their anus. This eliminated sugar is used by the ants, which even caress the aphids' abdomens with their antennae, causing them to eliminate more sugar. The ants carry the aphids back to their anthills and place them on delicate roots so that they can extract the elaborate sap from them. The ants often look after the aphids' offspring so that they can enslave them in the future (Figura 11), to obtain sugar. Some authors consider this type of interaction to be a form of protocooperation, particularly called synphily.

Figura 11 - Examples of slavery: ants and aphids.

4.2.3.5 Competition

This is the ecological interaction in which individuals of the same or different species compete for a resource (food, territory, light, water).

Competition can be intraspecific, when it occurs between individuals of the same species, or interspecific, between different species (CECIERJ, 2016). Competition induces a selective process that results in the preservation of life forms that are better adapted to the environment and the reduction of populations or extinction of less adapted individuals.

Competition is a factor that regulates the density of populations, helping to prevent the overpopulation of different species.

4.2.4 Demographics

The demography of populations studies their structure, including:
- the distribution by age,
- the birth rate,

> number of individuals per gender,
> dimensions,
> mortality rate,
> breeding sites,
> migration and
> geographical distribution.

The study of demography makes it possible to determine how wild animal populations are developing over time and to make predictions for the future, so it is carried out by monitoring to collect data over time that allows predictions to be made.

Population regulation is the process by which wildlife populations are maintained in numbers within sustainable limits in their habitats. Regulation can occur intrinsically (competition for resources, predation and shelter), or extrinsically (environment, disease and anthropism).

The geographical distribution of populations in habitats creates spatial patterns, which can vary between species and is influenced by factors such as the availability of resources, predation pressure, competition and environmental characteristics.

Some wild animals live in social groups, while others are more individual. There are others that form groups of females and cubs, while adult males live alone, only approaching females for reproduction (e.g. coatis).

Demographic aspects, growth, population regulation and spatial patterns help to develop management strategies to conserve wildlife populations, restore degraded habitats and minimize conflicts between animals and humans.

Continuous monitoring of wildlife populations is recommended in order to evaluate the results of conservation measures and adopt corrective and preventive actions in population management.

4.2.4.1 Natural selection and adaptation

Wild animals are permanently subjected to evolutionary pressures, adapting to the situations of each moment and environment, where the strongest and most adapted survive and the weakest and least adapted have their populations reduced or extinguished, what is understood as natural selection.

In the process of natural selection, favorable characteristics are transferred to the next generation, while unfavorable ones are eliminated, because the strongest and most adapted are dominant and have a greater chance of surviving and reproducing.

Adaptation occurs through changes in physical or behavioral characteristics that increase fitness for the environment, usually occurring over several subsequent generations and as a result of random mutations in individuals in the population, which when they imprint superior characteristics and are passed on to subsequent generations make the individuals stronger and more adapted, becoming dominant in the population.

Adaptations can manifest as improved mimicry and camouflage, physical structure, feeding behavior and reproductive strategies, among others, as a response to the selective pressures of the environment, such as water and food scarcity, intra- and interspecific competition and predators.

4.2.4.2 Bionomic strategies

Bionomics studies the behaviour of species in relation to their environment, the relationship between them and how they are organized.

Bionomic strategies allow animals to survive and thrive in inhospitable environments as a result of a process of adaptation, shaping their physical, behavioral and physiological characteristics in such a way as to enable them to survive and reproduce.

The main bionomic strategies include:
> Camouflage and mimicry - confusing themselves in the natural environment to escape predators or approach prey undetected.

- ➤ Migration - common among species of birds, fish and mammals that travel seasonally in search of food, ideal climatic conditions or breeding sites.
- ➤ Hibernation and aestivation - hibernation in the cold months, drastically reducing its metabolism to save energy, or aestivation during periods of extreme heat or drought.
- ➤ Cooperation and Social Behavior - life in social groups, such as coatis, elephants, bees and vinegar dogs, which facilitates hunting, protection from predators and the creation of a social structure that contributes to the reproduction and survival of the species.
- ➤ Varied diet - the type of diet is a vital strategy for the survival of many wild animals, varying with the time of year and local conditions.
- ➤ Strategic reproduction - the adoption of specific reproductive strategies to maximize the survival of offspring. Some turtles, for example, lay large numbers of eggs, while elephants have prolonged parental care to ensure the survival of their young.
- ➤ Resistance to poisons and toxins - resistance to poisons and toxins;
- ➤ Production of poisons, toxins and strong odor - poisons to immobilize or kill prey and toxins or odor to ward off predators.

4.2.4.3 Abundance and availability of resources

Trophic chains begin with the availability of minerals in the environment, water and sunlight used by plants. Today's species have adapted to different environments and have evolved to take advantage of available resources, protect themselves, shelter, feed and reproduce. Depending on the evolution they have undergone, each species has specific needs and, depending on whether the environment is more or less favorable to them, they have developed more or less numerous populations.

Environments with an abundance of resources have larger populations and generally higher species richness (e.g. the Amazon). It should be considered that seasonality in many regions implies seasons with a greater abundance of resources and a more favorable climate, which has led to migratory species taking advantage of more favorable seasons to feed in certain places and to reproduce in others, such as whales.

Many species have adapted to these seasonal changes with advantages for their survival.

The availability of resources influences the distribution, population density and survival of species in different natural ecosystems.

4.2.4.4 Patterns and dynamics of intra- and interspecific interactions

Wild animals have developed patterns and dynamics of intra- and interspecific interactions in different ecosystems. The behavior of animal species is influenced by their interactions with each other. Interactions between individuals of a species (intraspecific) and between different species (interspecific) play a fundamental role in ecology and in maintaining the balance of ecosystems.

Intraspecific and interspecific interactions are dynamic and are the result of different environmental factors, such as seasonal variations in resources, climate change and disturbances caused by natural or anthropogenic factors in ecosystems. The selective pressures resulting from these interactions induce the evolution of species over time. The main interactions between animals are described in section "4.2 - Ecological interactions".

4.2.4.5 Co-evolution

Coevolution is the process in which two or more species coexist and interact with each other, influencing each other's evolution.

Interactions between coevolving species can be predation, competition for resources, symbiosis and mutualism between species living together in a certain habitat.

4.2.5 Communities and Metacommunities

Communities are groups of different species that coexist in a given place at the same time. Wild species occupy specific ecological niches and species interact with each other.

The structure of a community, including its species richness, the relative abundance of each species and the interactions between them over time, leads to the stability and resilience of ecosystems. The greater the biodiversity, the more likely a community is to adapt to environmental changes and resist natural disasters.

Metacommunities are made up of multiple communities (patches) in an anthropized matrix. At least some of the animal and plant species migrate between the various natural communities that make up the metacommunity in search of food, breeding partners and suitable places to raise their offspring, or they simply disperse in search of new areas. Meta-communities can even evolve in a similar way to natural areas when most of the species manage to migrate between the various communities on the islands or natural patches that make up the area covered by the meta-community.

5 PROTECTION OF NATURAL AREAS

In 2020, more than 17% of terrestrial and inland water areas and 10% of coastal and marine areas, mainly areas of importance in relation to biological diversity and ecosystem services in Brazil, were conserved by effectively and equitably managed, ecologically representative and interconnected systems (UNEP-WCMC and IUCN, 2021).

5.1 First national parks

The history of protected natural areas is a fascinating journey that spans centuries and continents, highlighting the growing awareness of the importance of nature conservation. Two notable protected natural areas that have played significant roles in this history are Bogd Khan Uul National Park in Mongolia and Yellowstone National Park in the United States.

Bogd Khan Uul National Park, created by the Mongolian government in 1778, is located near the capital Ulaanbaatar and is a testimony to Mongolia's natural and cultural heritage, as well as being one of the oldest parks in the world. The area around the majestic Bogd Khan Uul mountain rises to 2,265 meters above sea level. It was chosen to be Mongolia's first protected area because of its scenic beauty and spiritual significance to the Mongols. Known as Holy Khan Mountain, it is considered sacred and has great importance in Mongolian culture. It served as a spiritual retreat for religious leaders. The Park covers some 416,200 hectares, with diverse ecosystems, from taiga to alpine forests, grasslands and wetlands, which are home to rich wildlife, including rare and endangered species such as the Mongolian wolf, snow leopard and Pallas's cat.

Yellowstone National Park was established in 1872 in the United States, in the state of Wyoming. It was created because of its natural beauty, its geothermal features, which include geysers and hot springs, its diversity of wildlife and to make it a preserved place for public visitation. Its creation activated a global movement to protect natural areas.

The history of these two parks stands out in the evolution of the concept of protected natural areas. In the case of Bogd Khan Uul, the initial emphasis was on the spiritual and cultural importance of the area. Yellowstone, on the other hand, laid the foundations for scientific conservation and nature tourism. The conservation movement has been growing around the world ever since, leading to the creation of a global network of protected natural areas designed to preserve biodiversity, natural beauty and ecosystem services for future generations.

In the 19th century, the industrial revolution took place. The growth of cities, industrialization and urbanization led to a growing disconnection between people and the natural environment. Pollution and the unbridled exploitation of natural resources led to environmental degradation and city life became associated with an existence far removed from nature. Life in the countryside came to be idealized as healthier and more balanced, a refuge from the bustle and pollution of the cities, influencing cultural and literary movements that celebrated nature and the simplicity of rural life.

In the 20th century, there was an increase in environmental awareness and perception of the damage caused by civilization. Pioneers such as John Muir in the United States and André Rebouças in Brazil defended the importance of preserving nature and establishing protected areas.

In the United States, the movement to create "natural areas" gained momentum with the creation of the National Park Service in 1916, under the leadership of Stephen Mather and Horace Albright, leading to the protection of natural landscapes of great scenic value, such as Yellowstone National Park, Grand Canyon National Park and many others. The first parks with natural monuments in the USA were created as a way of strengthening the country's identity. The land was being modified by human action and wild areas could be disfigured. The aim was to preserve wild areas without anthropism.

In Brazil, the creation of natural areas has also played a fundamental role in preserving biodiversity and ecosystems. Itatiaia National Park, founded in

1937, was Brazil's first national park and marked the beginning of an ongoing effort to protect areas of great environmental importance in the country.

5.2 Environmentalists

Engineer André Pinto Rebouças (1838-1898) was one of the pioneers in proposing the creation of National Parks in Brazil. Together with his brother Antônio, they built some of the most important structures in the southeast and south of the country, such as the Curitiba-Paranaguá railroad project and the Graciosa Road. Still in the 19th century, he suggested the creation of the Guaíra National Park in Paraná. Inspired by movements in the United States to create protected areas such as Yellowstone Park, André Rebouças, in a text published in the IHGPR bulletin in 1917, described the benefits of creating conservation areas for public visitation and environmental protection. According to him, the opening of these parks would benefit "future generations".

John Muir (1838-1914) is considered one of the founders of the modern conservation movement; he became known as "John of the Mountains" and "Father of the National Parks". He was a Scottish-American naturalist, writer, environmental philosopher, botanist, zoologist, glaciologist and advocate of wilderness preservation in the United States of America. Born in Scotland, he emigrated to the United States with his family as a child. Muir explored the wild lands of North America, from the Rocky Mountains to California, documenting in detail the diversity of natural environments, gaining credibility and respect and going on to influence the nation's mentality towards conservation. He played a key role in the creation of national parks in the United States, including Yosemite National Park in California, established on October 1, 1890.

In the 20th century, some prominent environmental activists included Jean Dorst, Rachel Carson, David Attenborough, Jane Goodall, Wangari Maathai, Jacques-Yves Cousteau, Dian Fossey, Aldo Leopold and Chico Mendes.

Jean Dorst (1924-2001) was a renowned French ornithologist and ecologist. He played a key role in creating protected areas and promoting the conservation of endangered species. He became known for his significant contributions to the study of ornithology, ecology and wildlife conservation. "Before Nature Dies" is a remarkable work by renowned French biologist and ecologist Jean Dorst, originally published in 1965 and continually relevant in our contemporary world. The book eloquently calls for action in favor of environmental conservation and the protection of biodiversity. Dorst provides an in-depth and insightful view of the problems and attacks on nature due to the growing influence of human activities. He shows the unbridled exploitation of natural resources, pollution, habitat destruction and threats to fauna and flora as direct consequences of an industrialized society eager for progress. The strength of Dorst's work lies in his ability to translate complex scientific issues for a wider audience. He presents clear and convincing evidence of the damage caused by humanity to nature and argues that if we don't take immediate action to curb this destruction, we will face serious consequences for the ecological balance of the planet. Throughout "Before Nature Dies", Jean Dorst also highlights the importance of education and public understanding of nature as a fundamental part of the solution. He believed that knowledge and environmental awareness are key to inspiring change and instigating action to preserve the natural environment. Although it was written decades ago, "Before Nature Dies" is a timeless work that reminds us of the urgency of protecting our planet and the diversity of life it harbors in order to prolong our survival. It is a call to reflection, to action and continues to inspire generations of environmentalists and nature defenders to work tirelessly for the conservation of our natural world.

Rachel Carson (1907-1964) was a marine biologist and renowned author. She is best known for her 1962 book Silent Spring, which alerted the public to the dangers of pesticides, especially DDT, to the environment and wildlife. Carson is often remembered for laying the foundations of the modern environmental movement.

Sir David Attenborough (born 1926) is a renowned British naturalist and documentary maker. He is famous for his wildlife documentaries and series, such as "Planet Earth" and "Life on Earth", which have educated and made millions of people aware of the beauty and fragility of our planet.

Dame Jane Morris Goodall (born 1934) is a British primatologist known for her pioneering study of chimpanzees in Tanzania. She is an advocate of wildlife conservation and the protection of natural habitats. She set up the Jane Goodall Foundation, which works to promote conservation and environmental education.

Wangari Muta Maathai (1940-2011) was a Kenyan teacher, environmentalist and political activist. She created the Green Belt Movement to promote tree planting and combat environmental degradation. She also fought for the empowerment of women in rural communities in Kenya. She was the first African woman to receive the Nobel Peace Prize in 2004.

Jacques-Yves Cousteau (1910-1997) was a French naval officer, oceanographer, documentary filmmaker and inventor. Among other inventions, he and Émile Gagnan invented the aqualung diving regulator, which replaced the heavy scapulars and allowed greater freedom and movement on dives. With the assistance of Jean Mollard, he made the SP-350 "Diving Saucer", an experimental diving vehicle that could reach a depth of 350 meters. He produced four feature films and around seventy television documentaries about nature and the oceans. With the film "The World of Silence", Cousteau won the Palme d'Or at the Cannes Film Festival in 1956, co-produced by Louis Malle.

Dian Fossey (1932-1985) was an American primatologist and conservationist known for her extensive study and protection of mountain gorillas in Rwanda. Her book "Gorillas in the Mist", published two years before her death, is an account of her work with gorillas at the Karisoke Research Center and her earlier career, and was made into a film of the same name in 1988.

Aldo Leopold (1887-1948) was an American forester, environmentalist and environmental philosopher and the founder of conservation science in the United States. He became known for his book "A Sand County Almanac" and for developing the land ethic, which emphasizes the need for a responsible relationship between people and the land they inhabit. Leopold profoundly influenced the development of environmental ethics. He participated in the founding of The Wilderness Society in 1935. He then acquired land in the Wisconsin countryside, where he used innovative ecological restoration practices, the experiences of which were collected posthumously in his work "A Sand County Almanac: And Sketches Here and There", published in 1949.

In Brazil, Francisco Alves Mendes Filho (1944-1988), known as Chico Mendes, was a Brazilian trade union leader and environmental activist. Mendes grew up in Acre and began his career as a rubber tapper. The devastating impacts of uncontrolled logging and deforestation in the region led him to get involved in organizing rubber tappers into unions to fight for their rights and protect the forest. He proposed the creation of extractive reserves. Chico Mendes' legacy continues to inspire the fight to conserve the rainforest and protect the rights of the communities that depend on it.

All these environmentalists and many others played decisive roles in raising awareness of nature conservation throughout the 20th century and their works continue to inspire action to conserve the natural environment.

5.3 First UN conferences on the environment

The United Nations Conference on Development and the Human Environment took place between June 5 and 16, 1972 in Stockholm, bringing together 113 countries. It was a historic milestone, being the first international meeting with representatives from different nations to address environmental problems. The result was the drafting of the Stockholm Declaration, with 26 principles, and the creation of the United Nations Environment Program (UNEP). The main issues addressed at the conference were air pollution, water and soil pollution and the consequences of population growth on natural resources.

In 1992, the United Nations Conference on Environment and Development took place in Rio de Janeiro, also known as ECO-92. It was the second United Nations world conference on the environment and brought together 178 countries. Some of the themes addressed at ECO 92 were: sustainable development, climate change, biodiversity and deforestation.

As a result of the Conference, the following documents were produced:

- Earth Charter - a declaration of fundamental ethical principles for building a just, sustainable and peaceful global society;
- Convention on Biological Diversity - deals with the protection of biodiversity;
- United Nations Convention to Combat Desertification - deals with the reduction of desertification;
- United Nations Framework Convention on Climate Change - deals with global climate change;
- Declaration of Principles on Forests - a non-legally binding document that makes a series of recommendations for conservation and sustainable forest development;
- Rio Declaration on Environment and Development - a United Nations proposal to promote sustainable development;
- Agenda 21 - a document that establishes the importance of each country committing itself and reflecting, globally and locally, on the way in which governments, companies, non-governmental organizations and all sectors of society could cooperate in the study of solutions to socio-environmental problems; it is a reinterpretation of the concept of progress, contemplating greater harmony and holistic balance between the whole and the parts, promoting quality, not just the speed of growth; the document guides each country to develop its own Agenda 21, being an instrument that guides the political construction of the foundations

of an action plan and participatory planning at global, national and local level, in a gradual and negotiated manner, with the goal of a new economic and civilizational paradigm.

6 THREATS AND EXTINCTION OF WILD ANIMALS

Extinction of a species is considered when there are no living individuals left.

Species of animals whose populations are reduced to the point where they could become extinct in the short to long term, under the environmental conditions existing at the time their populations were assessed, are considered to be at risk of extinction.

Extinction can be natural or man-made:

> ➢ Natural extinction is part of the evolution of species; those that are best adapted to the current environment survive and those that can't adapt become extinct;
> ➢ Anthropogenic extinction is caused by man.

The human population has been growing exponentially, reaching more than eight billion people. Consumption of natural resources is also growing exponentially and is close to exhaustion. This growth and development endangers the habitats and existence of various types of wildlife around the world, especially species used for food or other human purposes, or which lose their habitats due to occupation by the human population for rural production, transportation routes and urban areas.

Other threats to wildlife include the introduction of invasive species from other parts of the world, climate change, pollution, overfishing and hunting.

Graipel et al (2016) classified the risk of extinction of mammal species by the characteristics indicating susceptibility to extinction associated with biogeography (A-B), biology (C-E) and anthropogenic impacts (F), considering the following binary factors:

"**A)** Endemism: (0) non-endemic species; (1) species endemic to the Atlantic Forest;

B) Habitat: (0) species with non-restricted distribution; (1) species with restricted distribution, i.e. less than 50,000 km 2 ;

C) Mass: (0) small or medium-sized species, i.e. with a mass of less than 10kg; (1) large-sized species, i.e. with a mass of 10kg or more;

D) Diet: (0) non-carnivorous species; (1) carnivorous species, including strictly carnivorous, piscivorous or haematophagous;

E) Reproduction: (0) species with a high reproductive rate, with sexual maturity or independence reached in the first year for females; (1) species with a low reproductive rate, with sexual maturity or independence reached after the first year for females;

F) Hunting: (0) species whose hunting pressure poses no threat of extinction; (1) species whose hunting pressure poses a threat of extinction."

Salafsky et al (2008) suggested a standard vocabulary for biodiversity conservation with the aim of standardizing threat classifications and conservation actions and created two lists, the first on direct threats to biodiversity and the second on conservation actions. List of direct threats to biodiversity by the IUCN-CMP according to SALAFSKY et al (2008):

"1. residential and commercial development

2. Agriculture and aquaculture

3. Energy production and mining

4. Transport and service corridors

5. Use of biological resources

6. Human intrusions and disturbances

7. Modifications to the natural system

8. Invasive and other problematic species and genes

9. Pollution

10. Geological events

11. Climate change and severe weather"

Direct threats are generally limited to human activities that can be countered with appropriate actions (SALAFSKY et al, 2008). However, there is a fine line between a natural event, such as a fire caused by lightning and a fire caused by a phosphorus or the increased intensity of fires due to inadequate management of economic plant production areas). IUCN-CMP conservation actions according to SALAFSKY et al (2008):

"1. land/water protection

2. Land/water management

3. Species management

4. Education and awareness

5. Law and politics

6. Livelihoods, economic incentives and others

7. External training"

National and international organizations that care about and act to conserve the natural environment, such as WWF, CI, WCS and the UN work to:

➢ support global animal and habitat conservation efforts on many different fronts;

➢ with the government to establish and protect public lands, such as national parks and wildlife refuges;

➢ writing legislation to protect species;

➢ with law enforcement to prosecute wildlife crimes such as wildlife trafficking and poaching;

➢ conserve biodiversity to support the growing human population, preserving existing species and habitats;

➢ slow the extinction of global species and protect global biodiversity and habitats;

➢ for conservation efforts, documenting and drawing attention to endangered wildlife around the world.

6.1 Types of extinction

Three main types of extinction can be considered: mass extinctions, phyletic extinctions and background extinctions.

Mass extinction is the end of life for a large number of species in a short period of time. An example of this type of extinction was that of the giant mammals at the end of the last glaciation 10,000 years ago.

Phyletic extinction or speciation occurs through evolution and is known as pseudo extinction. In anagenesis, or phyletic speciation, a species undergoes modifications due to continuous environmental changes, until it gives rise to a different species and the original ceases to exist.

Background extinctions are the progressive and continuous disappearance of species over time due to maladaptation, competition between species, diseases such as that of *Atelopus zeteki, which became* extinct in the wild due to a fungus, or other natural causes.

The main threats imposed by humans on wildlife are:
- Habitat destruction, fragmentation and degradation
- Hit-and-run accidents on transport routes
- Exotic species
- Super exploration
- Pollution
- Climate change

Natural threats to wild animals can be:
- Geological disasters
- Space disasters
- Pests and diseases
- Endemism and low genetic diversity

6.2 Habitat alteration and landscape fragmentation

Humanity has been appropriating natural areas and altering habitats since prehistoric times, causing..:
- Elimination of natural habitats by:
  - mining,
  - construction of transport routes,

  - o buildings,
  - o dams and
  - o artificial lakes;
- ➢ Change of habitats to:
  - o agriculture,
  - o livestock,
  - o aquaculture and
  - o forestry;
- ➢ Habitat degradation by:
  - o abuse and
  - o pollution.

The reduction and degradation of natural habitats is one of the main causes of threats to animal species, which can lead to their extinction as a result of reduced population density, genetic diversity and geographical distribution.

6.2.1 Landscape

Landscape is the area of land that is covered by a single view. It is a limited terrestrial unit composed of the shape of the relief, soil and vegetation.

For Metzger (2001) landscape is "a heterogeneous mosaic made up of interacting units, this heterogeneity existing for at least one factor, according to an observer and at a given scale of observation".

The landscape is a basis for studying topological and chorological relationships (Figura 12) of landscape ecology. The more dynamic attributes of the land, such as certain animal populations and water flows, are less suitable as diagnostic criteria, but often link units by characteristic flows.

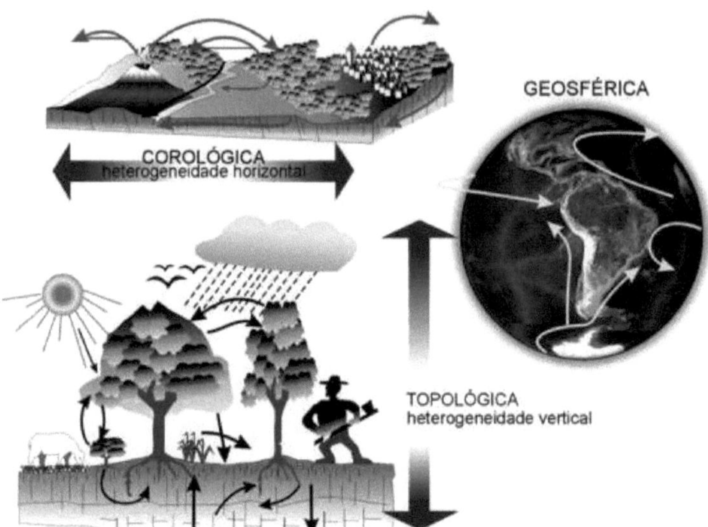

Figura 12 - Chorological and topological heterogeneities in landscape studies, according to Neef (1967). Vertical heterogeneity is due to the attributes of the place and horizontal heterogeneity to the relief units, which can be distinguished in a chorological classification at various scales, up to the geospheric scale. Source: Porto and Menegat (2015).

The landscape is dynamic, changing with the movement of continents, volcanism, climatic events and human action that causes the fragmentation of vegetation cover, defined as landscape fragmentation. The factors that shape the landscape are activated by time (Figura 13).

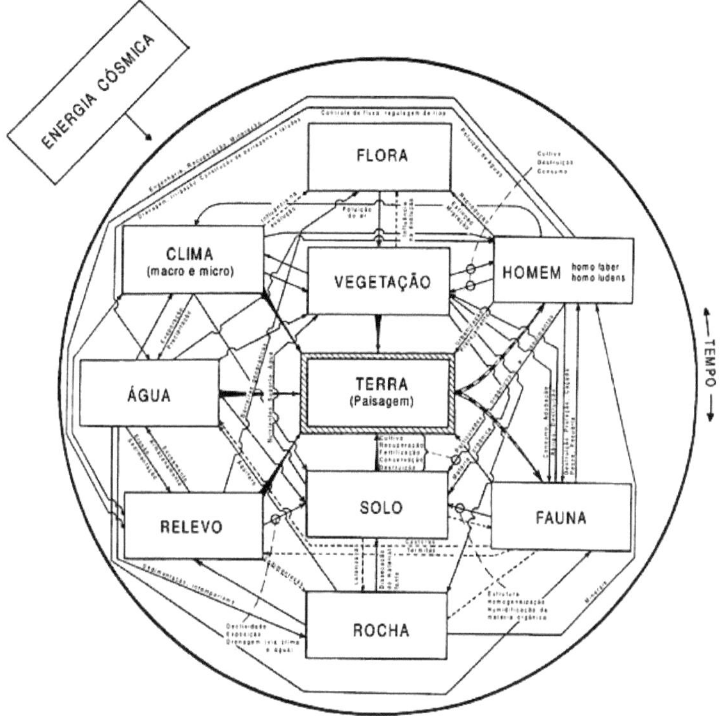

Figura 13 - Landscape formation factors, their attributes and relationships. Source: ZONNEVELD (1972) apud Soares Filho (1998).

6.2.2 Fragmentation of the landscape

Landscape fragmentation is the process by which an area of continuous natural habitat has most of its surface altered by human action to form an anthropized matrix, leaving two or more fragments of original habitats. Landscape fragmentation causes the isolation of portions of the land with original vegetation that may be home to animals.

A fragmented landscape is a heterogeneous territorial unit made up of a mosaic of interacting units or patches of different habitats on a dominant matrix. The fragmentation process can generate shrunken, bisected, fragmented or perforated habitat models (Figura 14).

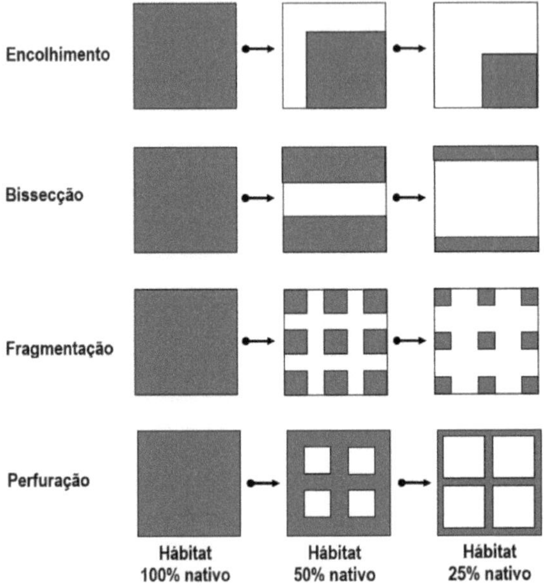

Figura 14 - Models of landscape fragmentation processes. Source: Collinge (2009).

6.2.3 Landscape ecology

According to TROLL (1968) *apud* Porto and Menegat (2015), landscape ecology is the total (integral) study of a given area, considering the complex effect between biocenoses and relationships with the environment, finding this organization and a certain distribution pattern in different orders of magnitude.

Landscape ecology can have two different approaches (METZGER, 2001):
> Geographical - the study of man's influence on the landscape and land management;
> Ecological - study of the effect of spatial context on ecological processes and their influence on biological conservation.

Concepts in the study of landscape ecology (METZGER, 2001):
> Edge - Transition area between two landscape units.

- Connectivity - The capacity of the landscape (or landscape units) to facilitate biological flows. Connectivity depends on the proximity of habitat elements, the density of corridors and stepping stones, and the permeability of the matrix.

- Corridors - Homogeneous areas (on a given scale) of a landscape unit that are distinct from neighboring units and have a linear spatial layout. In fragmentation studies, a corridor is only considered to be the linear elements that connect two previously connected fragments.

- Landscape element - This is each patch, corridor or area of the matrix. A landscape unit can have several elements in a landscape. For example, a "forest" unit can have several patches and some corridors.

- Scale of perception - Spatial and temporal scale at which each species perceives the landscape according to its ecological characteristics (size of territory, specificity of habitat, ability to move, etc.).

- Spatial scale - This is defined by the characteristics of extent (size) and resolution (minimum unit of spatial representation). Maps vary from point and fine scales (detailed maps, with high resolution and, in general, reduced extent) to global and coarse scales (maps with few details, with coarse resolution and, in general, wide extent).

- Fragment - A spot caused by fragmentation, i.e. by sub-division, promoted by man, of a unit that initially appeared in a continuous form, like a matrix.

- Patches - Homogeneous areas (at a given scale) of a landscape unit, which are distinguishable from neighboring units and have reduced and non-linear spatial extensions.

- Matrix - Territorial unit that controls the dynamics of the landscape (FORMAN, 1995 apud Metzger, 2001). In general, this unit can be recognized as covering the largest part of the landscape (i.e., being the dominant unit in terms of spatial coverage), or as having a greater degree of connection in its area (i.e., a lower degree of fragmentation). In a second definition, particularly used in fragmentation studies, the matrix is understood as the set of non-habitat units for a given community or species studied.
- Mosaic - A landscape that has a structure containing a patch, corridors and matrix (at least two of these elements).
- Landscape - A heterogeneous mosaic made up of interacting units, with this heterogeneity existing for at least one factor, according to one observer and at a given scale of observation. A landscape can be presented as a mosaic, containing patches, corridors and a matrix, or as a gradient.
- Fractal system - A system that maintains its characteristics and properties at different scales.
- *Stepping stone* - Small areas of habitat scattered throughout the matrix that can, for some species, facilitate flows between patches.
- Landscape unit - Each type of landscape component (land cover and land use units, ecosystems, vegetation types, for example).

The quality of life for species of fauna in fragments is lower the smaller the number of fragments, and the more distant and smaller they are. The lack of natural vegetation corridors between fragments hinders gene flow between them and can lead to inbreeding in some animal species due to the excessive isolation of small populations. These factors all constitute threats to the survival of fauna species in fragments.

The fragmentation process transforms a natural matrix into an anthropized matrix with sparse islands of natural vegetation (Figura 15).

Figura 15 - The process of fragmentation and the transformation of the matrix.

Each of the fragments or patches of original vegetation may or may not have a subpopulation of a particular species and the set of subpopulations in the entire area is called a metapopulation. The existence of a metapopulation can benefit from corridors (Figura 16) that allow the exchange of genes between subpopulations.

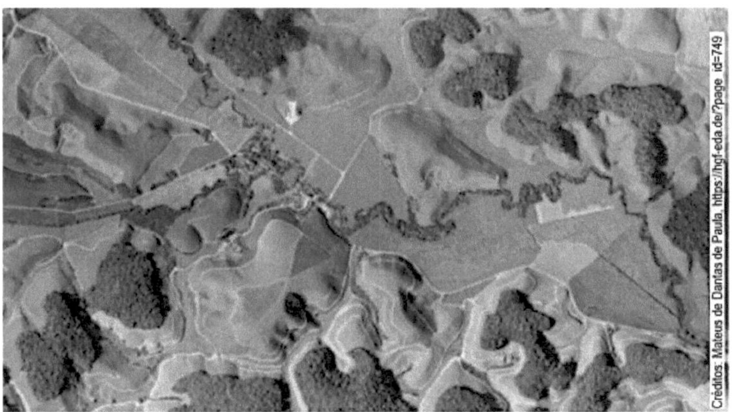

Figura 16 - A landscape composed of an anthropogenic matrix, fragments and corridors of natural vegetation, allowing the translocation of genes between the subpopulations that inhabit the different fragments.

The matrix of a fragmented landscape has various effects such as:
- ➢ Regulation of edge effects in fragments;
- ➢ It is a source of disturbance to the fragments;
- ➢ It acts as an alternative habitat for some species;
- ➢ Controlling flows between habitat fragments;
- ➢ Regulating the use of corridors and *stepping stones*;
- ➢ Regulates sensitivity to fragmentation.

Fragments, also known as patches or islands, can be natural habitat, introduced habitat or disturbed natural vegetation, and may offer some kind of resource necessary for the survival of wild animals. Disturbance patches are altered areas within a matrix made up of original vegetation (Figura 17) and represent around 7% of the national territory.

Figura 17 - Disturbance patches in a natural forest matrix.

The size, shape, ratio of perimeter to surface area of the patch, degree of isolation and time of disturbance are highly correlated with the degree of habitat disturbance caused by the edge effect.

Ecological corridors can reduce the negative effects of fragmentation by allowing the transmission of genes from one patch of natural vegetation to another and are considered fundamental for landscape connectivity. Among the functions of corridors are:

> Facilitate water and biological flows in the landscape;
> Reduce the risk of local extinction and encourage recolonization, thus increasing the survival of metapopulations;
> Landscape habitat supplement;
> Refuge for fauna when disturbances occur;

> Facilitate the spread of certain disturbances, such as fire or certain diseases;

> Reduced inbreeding in spots.

Connectivity represents the degree to which a landscape facilitates or impedes the movement of individuals and ecological processes between its different units, especially between patches, and is essential for maintaining healthy subpopulations. Connectivity can be understood in two different ways:

> Structural connectivity - This relates to the spatial arrangement of habitat units in terms of variation in size, degree of isolation, density and complexity of movement corridors; it is the physical continuity of a certain environment, including the existence of corridors and connection points and the distance (isolation) between patches of the same type of landscape unit;

> Functional connectivity - refers to the intrinsic ability of ecological processes or individuals of each species to move through the landscape mosaic, and this ability is species-specific and depends on morphophysiological patterns and their relationship with the different types of existing environments; it can be assessed by the ability of species to move through the matrix; the concept is closely linked to the difficulty of transit of individuals or processes imposed by the type of environment.

The structure of the landscape is represented by the following features:

> Area (size) - relates to the capacity of the habitat unit to maintain species;

> Spatial arrangement - relates to the habitat unit's capacity to receive immigrants;

> Shape - is related to the ability of the habitat unit to minimize the edge effect.

6.2.4 Consequences of habitat fragmentation

The main consequences of fragmentation are as follows:

> Reduction in habitat area and consequent reduction in resources for fauna;
> Changes in the spatial arrangement of habitat units and consequent greater isolation of subpopulations;
> Increased edge effect with consequent invasion of species from the matrix and also reduction of species from the patch, altering the species composition.

6.3 Island biogeography

The similarities between the theory of island biogeography (TBI) and the theory of metapopulations (TM) have been considered by many. Both focus on immigration and extinction in patchy landscapes and both are relevant to understanding populations and communities in fragmented landscapes. IBT emphasizes species richness on islands, while TM focuses on the persistence of single-species populations. But they also differ because in IBT there is a continent that serves as a source of colonization for new areas, while in TM, this may or may not be the case (it only exists for metapopulations between continent and island in which the continent is the disperser). In classical metapopulation theory, patches do not differ in isolation either, as they do in IBT. The theories are similar in the sense that both IBT and TM are equally nebulous about the variation in habitat quality between patches or islands, as well as about the characteristics of the matrix that can affect colonization and extinction rates, with the matrix as an essentially undefined backdrop.

Olof Vilhelm Arrhenius is remembered for the curve known as the "Species-Area Ratio" (SAR), the relationship between the number of species present and the size of the area hosting them, which for years was considered "one of the few genuine laws of ecology". Other researchers have observed the same phenomenon. In all cases, a potential positive relationship between area and species richness was observed, as described below:

$$S = c\,A^z$$

Where: S = number of species estimated for the island; A = area of the island; c, z = coefficients adjusted to the data.

6.4 Mining

Ores have been essential resources for civilization since the age of metals. They are used in all human activities, from mining for landfill and paving materials, building construction, cooking food, to the production of art objects and jewelry.

Mining is responsible for altering the landscape and contaminating the soil, air and water resources, reducing populations of all wildlife species where it is practiced.

6.5 Transport routes

Transportation routes, whether for pedestrians, cyclists or motor vehicles, involve changes to the landscape, the use of mined materials in borrow areas (Figura 18), removal of unsuitable natural material from the roadbed, which is disposed of in dump areas and cleaning of the road's domain area.

Figura 18 - Gravel extraction in a borrow area. Source: botucatuambiental.com.br

In addition to the destruction of the ecosystems through which the road passes, the borrow areas and the dump sites (Figura 19), the movement of vehicles involves running over animals crossing the road, which is one of the biggest causes of wildlife mortality.

Figura 19 - Disposal area for inorganic waste. Source: photo Tiel jornaldosudoeste.com.br

The signage (Figura 20), regarding animal crossings, has proven to be ineffective and there are almost no structures such as fences to prevent animals from entering the roadway domain area, bridges and tunnels to allow them to cross safely.

Figura 20 - Signaling animals crossing the road. Sources: Correio Braziliense (09/08/2020) and Curitiba City Hall (12/11/2021).

During the period of use and maintenance of the roads, there is air pollution caused by the discharge and dust caused by vehicles, water pollution caused by the carrying of debris disintegrated from the roadbed and erosion of the roadbed, cut slopes and embankments (Figura 21) and culvert outflows (Figura 22), as well as siltation of bodies of water that alter aquatic environments.

Figura 21 - Erosion on a slope on the side of a highway. Source: sistemavetiver.blogspot.com

Figura 22 - Erosion in culvert. Source: g1.globo.com

All the problems pointed out with regard to transport routes can be reduced with better planning and proper maintenance.

Construction of wildlife fences, bridges and tunnels allow animals to cross safely. Speed cameras have been used to reduce speeds on some stretches where wildlife crosses frequently.

Hit-and-run accidents cause death (Figura 23) and, worse, suffering to the animals that are run over and don't die instantly.

Figura 23 - Hit-and-run accidents are one of the biggest causes of the reduction in animal biodiversity.

Article 32 of Law 9.605/98 states that it is a crime to "abuse, mistreat, injure or mutilate wild, domestic or domesticated, native or exotic animals", which carries a penalty of three months to a year in prison and a fine, but there is no record of anyone having been convicted of running over wild animals.

475 million wild animals are killed on Brazilian roads (Figure 25), representing 17 deaths per second and 1.3 million per day (KAFRUNI and SOUZA, 2020). Many accidents involving animals could have been avoided with works to prevent them from entering the highway right-of-way and structures for them to cross (Figura 24).

Figura 24 - Structures that allow wild animals to cross roads safely. Source: Guimarães (2015).

Not just roads, but trains (Figura 25) and ships (Figura 27) also cause wildlife to be run over, injuring and killing animals.

Figura 25 - Running over wild animals on railroads.

A Figura 26 shows statistics on animals run over on highways.

Figura 26 - Results of animal traffic on highways. Source: Kafruni and Souza, 2020.

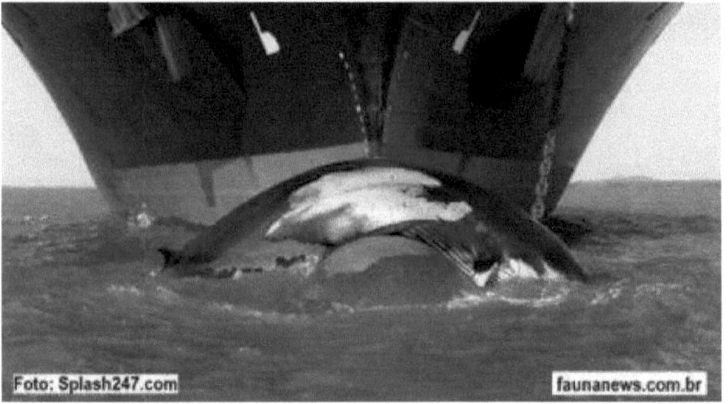

Figura 27 - Whale hit by oil tanker.

And although there aren't many reports in the media, aircraft collisions with birds are also common (Figura 28). Brazil recorded 4,600 reports of incidents involving aircraft and animals in 2013. CENIPA is the body responsible for recording and managing aircraft accidents in Brazil and reported the following:

> "The Center for Investigation and Prevention of Aeronautical Accidents (Cenipa) published on Tuesday (01/07) the annual Brazilian Fauna Risk data for 2013. The Aeronautics agency received 4,600 reports of incidents involving aircraft and animals at airports in Brazil. Most of the reports were sightings, with 2,344 cases.
>
> The 2013 statistical overview shows that the quer-quero, caracara and burrowing owl are among the bird species that have been most involved in collisions with aircraft.
>
> Often the plane takes off and the crew doesn't notice the collision. On the other hand, the ground crew identifies the dead animal, but can't pinpoint which aircraft collided with it, let alone which part was hit.
>
> The majority of collisions occurred during take-offs (25%) and landings (26%). This means that it is necessary to increase fauna dispersal activities to reduce the presence of animals on the runway when the aerodrome is in operation. (POGGIO, 2014)"

Figura 28 - Aircraft collisions with birds.

In Brazil, some rare initiatives have been created for wildlife crossing highways (Figura 29) and railroads (Figura 30).

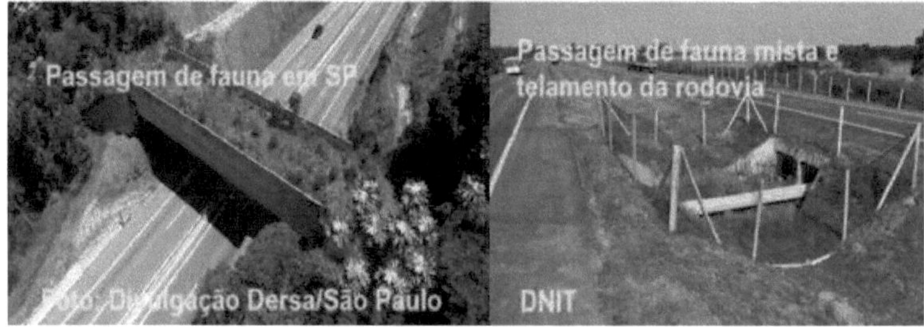

Figura 29 - Structures to protect wildlife from being run over on highways. Source: DNIT, 2014.

Figura 30 - Bridges and fences to protect wildlife on railroads.

6.6 Invasive species

"Invasive alien species are considered the second biggest cause of species extinction on the planet and can cause damage to biodiversity, the economy and human health" (IBAMA, 2022a). These are organisms that, when introduced outside their natural distribution area, threaten biological diversity and ecosystem services. The absence of natural predators, the abundance of prey without efficient natural defenses against the introduced species and disturbances in natural areas often create advantages for invasive alien species over native species. Invasive species are considered one of the biggest causes of species extinction on the planet, directly affecting biodiversity, the economy and human health (IBAMA, 2022a).

A species can become invasive when one or more of the following characteristics and conditions are present: rapid individual growth and maturation, high fecundity, capacity for rapid population growth, broad environmental tolerance and adaptability, absence of predators or antagonistic species and abiotic factors suitable for the invasive species. The greater the invasion potential, the more of these characteristics are present.

Invasive plants compete with native species for space or prevent native plants from growing, reducing the diversity of local flora and even fauna, as they often don't provide food or are toxic.

Invasive animals compete with wild species for space and food. They also transmit diseases to other animal species and to humans, cause agricultural damage by destroying crops and pastures, cause erosion and destruction of springs and native vegetation, and prey on native animals.

The Implementation Plan for the National Strategy for Invasive Alien Species aims to achieve the objectives and result indicators defined in the National Strategy and was established by SBio/MMA Ordinance No. 3/2018.

The Ministry of the Environment (MMA) has listed the invasive alien species sun coral (*Tubastraea* spp.), wild boar (*Sus scrofa)* and golden mussel (*Limnoperna fortunei*) as priorities for drawing up and implementing National

Prevention, Control and Monitoring Plans, and the Federal Government has included the three invasive alien species in the 2016-2019 Multi-Year Plan, with the aim of controlling and mitigating their impact on Brazilian biodiversity.

In Brazil, 365 exotic species are considered invasive, of which 197 are animals and 168 are plants (Figura 31).

6.6.1 Wild boar - *Sus scrofa* (Linnaeus)

The European wild boar is an artiodactyl from the *Suidae* family that can live up to 27 years. Adult males can weigh between 50 and 350 kg and females between 40 and 200 kg. It is native to Europe, Asia, the Sound Islands and North Africa (IBAMA, 2023).

Figura 31 - Preliminary analysis prepared by the Species Conservation and Management Department of the Ministry of the Environment (MMA, 2019).

Adult female wild boars live in packs with their young and cubs, made up of up to hundreds of individuals, while adult males are solitary. It has both nocturnal and diurnal habits. It is not very habitat-selective, living in woods and forests, but also frequenting open areas and cultivated areas at night.

It has a high reproductive rate in favorable conditions and can have up to three litters a year, with six to ten young. They live up to 10 years old in the wild and have no predators in Brazil. They are omnivorous animals, feeding on roots, fruits, seeds, berries, leaves, shoots, bulbs, small animals, fungi and carrion.

Your r

The introduction of the European wild boar to Brazil is controversial. According to Rosa et al (2018) it occurred as follows:

> "The aselvajada forms of *Sus scrofa* established their first large populations after entering central-western Brazil, probably due to the release of individuals at the time of the Paraguayan War, more than 200 years ago (Desbiez et al. 2011). In this region, they are currently known as "pig-monteiro". The second episode of introduction was officially reported in 1989, in Rio Grande do Sul, in a region close to the border with Uruguay, for commercial breeding purposes (Deberdt & Scherer 2007). Manipulated and accidental crossbreeding between the domestic pig and the wild boar resulted in a third morphotype, popularly called the "javaporco". Between 2000 and 2005, there were deliberate and widespread releases of wild boars and javaporcos due to fear of environmental inspection, especially in the south and southeast of the country, after the ban on new farms breeding the species (Ibama Ordinance No. 102/98) (Oliveira 2012)."

Figura 32 - Wild boars on the left and javaporcos on the right.

The wild boar is a hybrid of the domestic pig (*Sus scrofa domesticus*) and the European wild boar (*Sus scrofa scrofa*), and many populations of exotic wild pigs are made up of wild boars. The wild boar is smaller, with more developed canines that curve out of the mouth, a snout and a longer tail than the wild boar.

6.6.2 Sun coral - *Tubastraea* spp.

Tubastraea is a genus of corals that inhabit tropical reefs and rocky shores in shallow waters, from 0.5 to 15 meters deep. It originated in the Pacific and Indian Oceans, but has been distributed by ship hulls to almost all tropical regions. Two species occur on the Brazilian coast: *Tubastraea tagusensis* (Figura 33) (yellow color) and *Tubastraea coccinea* (Figura 34) (red-orange color) (IBAMA, 2022b).

Figura 33 - *Tubastraea tagusensis.*

Figura 34 - *Tubastraea coccinea.*

The sun coral has been observed in the Brazilian coastal zone (Figura 35), occurring in both natural and artificial environments, such as piers, buoys and oil platforms, with different stages of invasion and adaptation at each invaded site.

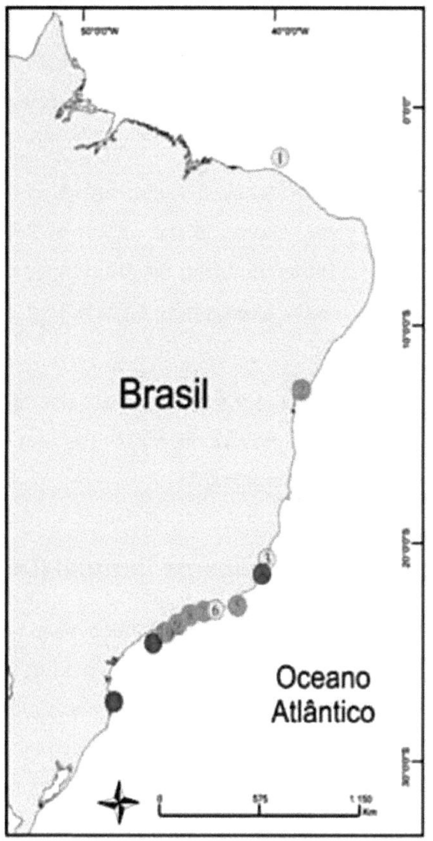

Figura 35 - Map of the occurrence of the sun coral in the different Brazilian states. Red circles: *Tubastraea coccinea*; Yellow circles: *T. tagusensis*; Green circles: T. coccinea and T. tagusensis. 1 - Acaraú (CE); 2 - BTS (BA); 3 - Vitória (ES); 4 - Guarapari (ES); 5 - Região dos Lagos (RJ); 6 - Cagarras (RJ); 7 - Baía de Sepetiba (RJ); 8 - Baía da Ilha Grande (RJ); 9 - Ilhabela (SP); 10 - Alcatrazes (SP); 11 - Laje de Santos (SP); 12 - Arvoredo (SC). Source: ICMBio (2018).

The sun coral was introduced to Brazil in the 1980s. The genus was first recorded on oil platforms in the Campos Basin. The first record on a rocky coast occurred in 1998, in Arraial do Cabo. Currently, there are records on the southeast and south coasts on natural rocky shores and artificial structures, as well as some records on the northeast coast, often associated with oil platforms.

Both species proliferate rapidly and invade areas where native corals occur, in some cases competing with and overlapping them.

The sun coral has been identified as an agent of modification of the benthic communities of rocky shores in the region of Ilha Grande, Rio de Janeiro, and coral reefs in Bahia, reducing the abundance of macroalgae and the cover of the native corals *Mussismilia hispida* and *Madracis decactis* (IBAMA, 2022b).

In June 2018, the Ministry of the Environment established the National Plan for the Prevention, Control and Monitoring of the Sun Coral (*Tubastraea* spp.) in Brazil, which was revised in 2020.

6.6.3 Golden mussel - *Limnoperna fortunei* (Dunker)

Limnoperna fortunei is a bivalve mollusc that lives in fresh and salt water (Figura 36). The species arrived in South America from Argentina and has spread throughout most of the South and several parts of the Southeast and Midwest. It is thought to have been introduced accidentally, in the ballast water of cargo ships (IBAMA, 2022c).

Although the larvae can swim, they can be carried freely by water or by vectors to solid surfaces, where they settle and grow to form large colonies. Adults are dispersed by transportation on boat hulls, nets, shells, branches and other objects in the water. Adults can survive out of the water, with their shells closed, for a long period of time.

Among the damage caused by the golden mussel are: destruction of aquatic vegetation; occupation of space and competition for food with native molluscs; damage to fishing, since the decrease in native molluscs reduces fish food; clogging of water, sewage and irrigation pipes and ducts; clogging of water intake systems for electricity generation, causing frequent interruptions for cleaning and making production more expensive; damage to navigation, with the compromise of buoys, traps, engines and boat structures (IBAMA, 2022c).

Eradication after infestation is practically impossible.

Figura 36 - Golden mussel - *Limnoperna fortunei*. Source: ima.sc.gov.br.

6.6.4 African snail - Achatina fulica (Ferussac)

Achatina fulica is a mollusc that originated in East Africa and was introduced to Brazil in the 1980s as a new source of food production.

Figura 37 - *Achatina fulica* (Ferussac).

This snail competes with native species for food and space, causes agricultural damage and transmits diseases to other animal species, including humans (JOLY et al, 2019). It is considered one of the worst invasive species, having been found in 23 Brazilian states (PARANÁ, 2021). It is most abundant in areas with a hot and humid climate.

Adults have an oval-conical shell with a tapered apex, brown in color with vertical, irregular and clear spots, 7-12 cm long and weighing around 100 grams or even more. It has a nocturnal habit and shelters from sunlight and heat during the day, remaining at rest attached to the hard vertical surface. It feeds on vegetables and organic waste such as food scraps, animal feed, feces, paper and organic waste in general. It is hermaphroditic, but rarely self-fertilizes. Fertilization can occur all year round in warm and humid regions and, in colder climates, in the hottest and wettest months. About 15 days after mating, it lays 10 to 400 eggs in the soil, partially or completely buried. When it reaches youth, it becomes more resistant to environmental adversities and can reach five years of age. Under very adverse conditions, it closes the opening of its shell with whitish, dry mucus, entering aestivation with reduced metabolism, and can remain so for months. The reduction in biodiversity is its greatest environmental impact and no predator of the species has been identified in the country. It transmits the nematode *Angiostrongylus cantonensis*, the agent of eosinophilic meningitis, and *Angiostrongylus*

88

costaricensis, the agent of abdominal angiostrongyliasis, both of which have few cases in Brazil (PARANÁ, 2021).

6.6.5 Lionfish - *Pterois volitans* (Linnaeus)

The lionfish originated in the Indian and Pacific Oceans. At the beginning of the 20th century, the species *Pterois volitans* and *Pterois miles* were already invasive on the east coast of the United States and arrived in the Caribbean and Gulf of Mexico in 2010. *P. volitans has been* found in Brazil since 2014. It inhabits coral reefs from 2 to 300 meters deep, tolerates turbid waters with varying pH and temperature, has good swimming ability, a long larval period and reaches a maximum length of 47 cm. They have dorsal spines that inject a toxin capable of causing nausea, pain and convulsions in humans. They proliferate rapidly and can lay up to 30,000 eggs. They are generalist predators, feeding on various organisms in the reef environment and have no natural predators in Brazil (ICMBio, 2023b).

Figura 38 - Lionfish - *Pterois volitans*. Source: Horus Institute.

The Lion Fish produces the following impacts in the areas it invades ICMBio (2023a):

> ➢ Predation on native and endemic species;
> ➢ Decrease in fishing production;
> ➢ Reduction of important species for the reef environment;
> ➢ Risk to human health.

6.6.6 Starling - *Sturnus vulgaris* (Linnaeus)

News published on the IBAMA website (2021):

> "Brasília (16/12/2021) - The Brazilian Institute for the Environment and Renewable Natural Resources (IBAMA) certifies the presence in Brazil of the bird of the species Sturnus vulgaris, known as the "starling". The first record of the bird in the country, considered to be invasive, occurred in 2014 - currently, its presence is restricted to municipalities bordering Uruguay."

The starling is a gregarious bird that originated in western and southern Europe and southwest Asia. It reaches a length of around 20 cm and a weight of between 70 and 100 g. It builds disorganized nests in hollows in trees and cliffs, where it lays four or five pale blue eggs that take two weeks to hatch; the chicks leave the nest after three weeks. They can lay up to two eggs a year. It is an omnivorous species, feeding on invertebrates, seeds and fruit, the eggs of other birds, with insects as its main diet.

Figura 39 - Starlings (*Sturnus vulgaris*). Source: IBAMA (2021).

It is considered a pest, reducing grain and fruit harvests and competing for food and space with native species, as well as transmitting ecto- and endoparasites and diseases (salmonella and toxoplasmosis) (BUSINARI, 2021).

6.6.7 Other invasive species

IBAMA (2023) lists other invasive alien species that cause considerable environmental impacts when in the natural environment, such as:

> *Felis catus* - domestic cat;
> *Mus musculus* - common mouse;
> *Oreochromis mossambicus* - Mozambique tilapia;
> *Lithobates catesbeianus* (= *Rana catesbeiana*) - Bullfrog;
> *Canis familiaris* - Domestic dog;
> *Callithrix jacchus* - White tufted marmoset;
> *Apis mellifera scutellata* - Africanized honeybee.

6.7 Overexploitation

The overexploitation of wild animal populations has caused the extinction of many species and still causes concern in the world today.

6.7.1 Predatory hunting

Animals are hunted because they pose a risk to humans and their livestock, as well as because they cause agricultural damage, for sport, or for food and trade. Alligators (Figura 40) and jaguars (Figura 41) are slaughtered for sport and to sell their skins, which has led many animals to drastically reduce their populations and even become extinct.

Figura 40 - Yellow-bellied caiman armor learned by the Environmental Squad in Boquim (SE). Source: G1 (18/04/2018).

Figura 41 - PF's Operation Jaguar seized Argentines, Paraguayans and Brazilians taking part in safaris in MS, MT and PR. Source: Gazeta do Povo (30/12/2023).

6.7.2 Wildlife trafficking

Security agencies, mainly IBAMA and the environmental police, have seized thousands of wild animals caught for trade, mainly birds (Figura 42), small primates and ornamental fish. The impact on their populations is enormous, many have been driven to the brink of extinction and some have become extinct.

Figura 42 - Birds seized by IBAMA in a scheme to defraud the Amateur Bird Breeding Activity Control and Monitoring System. Source: IBAMA (24/01/2020).

The illegal trade in wild animals generates between 10 and 20 billion dollars a year worldwide, of which approximately 10% corresponds to Brazil, with around 38 million wild animals removed from Brazil every year (MENUZZI et al, 2020). Animal trafficking is carried out for different purposes, as shown in Figura 43.

6.7.3 Sport, food and supplies

Many wild animals are slaughtered for sport and food, or as a source of inputs, such as armadillos, partridges, macucos, pacas and tapirs, in such numbers that many have been driven to the brink of extinction, or extinct. Turtles were killed for their meat and shells and their eggs were collected for food. Overfishing has been identified as the main cause of the reduction in fish schools and the average size of individuals. Whales were slaughtered to use their blubber as a source of energy for lighting and mortar for buildings, as well as their meat as food. Bones, ivory and leather also served as justification for the slaughter of wild animals, as did oils and other animal parts used in the production of perfumes and medicines.

Figura 43 - Destinations of wild animals traded. Source: Menuzzi et al (2020)

6.8 Pollution

Jeans Dorst's *Avant que nature meure*, published in 1971 and translated into Portuguese in 1973, influenced many scientists who saw the problems caused by the disruption of the main planetary balances and, after its publication, many organizations were created around the world to protect nature (ALMEIDA, 2008). In his book, Dorst (1973) denounced the main problems caused by the pollution and abuse of the natural environment by civilization and listed the main forms of pollution caused by man as:

" 1 - FRESHWATER POLLUTION

CHEMICAL POLLUTION WITH BRUTAL HARMFUL EFFECTS

Pollutants: toxic mineral products (mineral salts - heavy metal salts, acids, alkalis, ... - or organic - phenols, hydrocarbons, detergents, ...)

Responsible: all industries, due to massive accidental dumping.

CHRONIC CHEMICAL POLLUTION

Pollutants: phenols, hydrocarbons, various industrial wastes. Phytosanitary products (insecticides and herbicides). Synthetic detergents. Synthetic fertilizers (nitrates).

Responsible: various industries (refineries, oil, plastics, rubber industries: gas plants, coal plants, wood distillation, tars...). Agriculture. Domestic use and detergent industries.

BIOLOGICAL POLLUTION

Pollutants: fermentable organic waste.

Responsible: sewage from urban communities. Cellulose industries (sawmills, paper mills), textile and food industries (distilleries, breweries, canning factories, dairy industry, sugar industry, slaughterhouses). Tanneries.

PHYSICAL POLLUTION

Radiation pollution

Pollutants: radioactive waste from nuclear explosions and controlled nuclear reactions. Induced radioactivity.

Responsible: nuclear industries.

Mechanical pollution

Pollutants: inert solid materials (sludge, clay, slag, dust)

Responsible: large construction sites, road building. Extraction industry. Ore washing. Dredging.

Thermal pollution

Pollutants: dumping of cooling water that raises the temperature of rivers.

Responsible: thermal and nuclear power stations. Refineries. Various industries.

2 - POLLUTION OF THE SEAS

...waste... industrial waste... domestic and industrial waste water... hydrocarbons (oil and shipping waste)...

3 - AIR POLLUTION

...industries release an unsuspected amount of gases and solid waste into the atmosphere in the form of fine particles capable of remaining in suspension and passing into the airways of humans and animals, or of settling after having been transported, sometimes over considerable distances.

...

The same goes for pollution caused by cars.

4 - DISTURBANCES IN THE BALANCE OF THE ATMOSPHERE

Certain forms of pollution can have consequences for the entire planet, threatening the balance of the biosphere as a whole. This is the case with the production of carbon dioxide from the use of fossil fuels, coal, oil and natural gases.

...

The various pollutions for which man is responsible can also cause profound disturbances in climatic conditions.

...

5 - RADIOACTIVE POLLUTION

...

There are three types of activities responsible for radioactive pollution, affecting air, soil, fresh and salt water. ... atomic explosions ... the use of water in atomic power stations ... atomic waste... (DORST, 1973)."

As you can see, Dorst was already warning about many of the pollution problems, such as the spillage of crude oil (petroleum) as a result of the sinking of the Torrey Canion in 1967 (Figura 44), directly decimating bird populations and marine invertebrates in the south of England, indirectly affecting all food chains in the region (DORST, 1973).

Figura 44 - Wreck of the Torrey Canion on March 18, 1967. Source: Mesquita (2019).

Since then, the problems have worsened with the intensive use of plastics, mainly as packaging, and the excess waste generated by human beings. Civilization generates an absurd amount of sewage, organic and inorganic waste and rubble, which is disposed of with or without treatment in

97

bodies of water, dumps and landfills, much of which is carried away by rainwater, contaminating rivers, lakes and the sea, as well as the atmosphere while the organic waste decomposes. All these problems are still solvable, but they need effective and urgent action if they are not to reach the point of no return.

The main damage caused by sea pollution is as follows (LEITE, 2023):
> Dirty beaches unsuitable for leisure and fishing;
> Contamination of water, mainly with the presence of fecal coliforms from untreated sewage, as well as other harmful bacteria, harming animals and human beings;
> Occurrences of chemical waste, such as chlorine, mercury, chromium and lead, which contaminate fish, the environment and people;
> Reduction of fishing and farming environments for traditional communities;
> Impact on estuarine regions, mangroves, corals and bays - breeding grounds for hundreds of species of marine fauna. This is precisely where the greatest effects of pollution on the seas are to be found, with the dumping of waste and sewage from nearby towns, agriculture and industry.

6.8.1 1Fertilizers

Fertilizers are essential in food production to supply the world's population, but they pose threats such as contamination of the air, soil, rivers, lakes and oceans, imbalance in ecosystems and loss of biodiversity.

Nitrogen fertilizers contaminate the air through the evaporation of substances such as nitrous oxide, which is harmful to the ozone layer and contributes to global warming (CIVITEREZA, 2021). Together with potash fertilizers, they leach into the groundwater. It is also carried by rainwater into rivers, lakes and oceans, along with phosphates. Chemical fertilizers, especially phosphate and nitrogen fertilizers, eliminate soil microorganisms, resulting in soil impoverishment and hindering plant growth (CIVITEREZA, 2021).

In bodies of water, they cause eutrophication, causing an exponential increase in algae and aquatic plants. The excess of algae can reduce the availability of dissolved oxygen in the water, which causes the death of fish, corals and other organisms. The eutrophication of waters can promote the proliferation of toxic cyanobacteria, which can also cause the death of aquatic organisms.

The use of organo-mineral fertilizers considerably reduces, but does not completely avoid, the negative impacts of chemical fertilizers.

6.8.2 Pesticides

Rachel Carson's "Silent Spring" (1962) was a landmark in denouncing the negative impacts of pesticides on the environment, particularly on birds. Carson exposed the alteration of cellular processes in plants, animals and humans caused by pesticides. She publicized the decrease in the thickness of eggshells caused by DDT, leading to its banning in Switzerland in 1969 and in the United States in 1972.

Organochlorines such as DDT, Dodecachlor and BHC were widely used to combat arthropods; many were banned in 1985 by Ministry of Agriculture Ordinance No. 329, but Dodecachlor remained in use in Brazil until 1992, due to the lack of an efficient formicidal active ingredient to replace it at the time. These substances persist for up to 30 years in the environment and are cumulative in the organisms of animals that consume products contaminated with them.

Nowadays, different substances are used in rural production, such as: insecticides, fungicides, herbicides, nematicides, acaricides, rodenticides, molluscicides, formicides and growth regulators.

Agricultural crops require the use of pesticides to reduce production losses, but they also affect non-target organisms, such as large animals, mainly predators, by accumulating in their tissues; they also affect organisms in the soil, pollinators and aquatic ecosystems, which can cause environmental imbalance and problems for animal and human health (BELCHIOR et al, 2014).

More than 2.6 billion kilograms of pesticides are applied worldwide every year (Figura 45) and are responsible for pollution that affects the entire environment, reducing biodiversity and damaging the food webs of aquatic and terrestrial ecosystems.

The accumulation of pesticides in food chains is one of the biggest problems, poisoning top predators such as dolphins and other aquatic mammals, reducing their immunity, causing disease and even death, including humans (Figura 46).

One of the problems with pesticide application is wind drift, which carries the poison far and wide, affecting wildlife around crops, killing bees in apiaries, insects that serve as food for wild animals and killing native plant species that feed herbivores.

Aquatic plants, such as diatom algae, which are at the base of the food chain, when exposed to toxic chemicals have their tissues damaged, blocking the process of photosynthesis and causing an imbalance in the ecosystem.

Pesticides are associated with health problems such as chromosomal alterations, various types of cancer, respiratory diseases, among others (SOUSA, 2024).

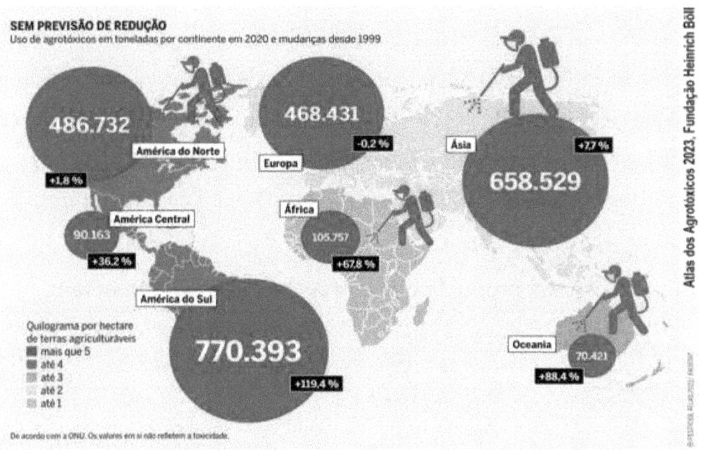

Figura 45 - Use of pesticides by continent. Source: TYGEL et al. 2023.

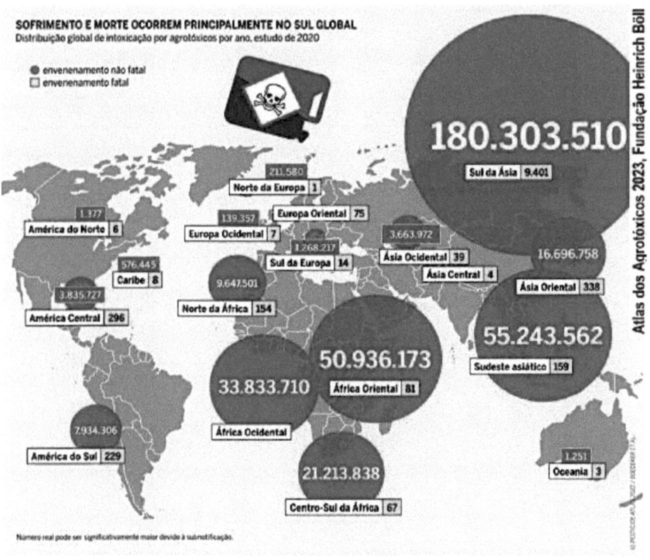

Figura 46 - Pesticide poisonings per year. Source: TYGEL et al. 2023.

6.8.3 Plastics

The world has already produced 10.1 billion tons of plastic, more than half in the last 20 years (GASINE, 2022). Brazil generated more than 11 million tons a year and recycled less than 1.3% in 2019 (Tabela 3). The amount of plastic in the oceans, for example, already exceeds 1.1 million tons (ERIKSEN, 2023).

TABELA 3 - Plastic production and recycling in the world in tons.

País	Total de lixo plástico gerado*	Total incinerado	Total reciclado	Relação produção e reciclagem
Estados Unidos	70.782.577	9.060.170	24.490.772	34,60%
China	54.740.659	11.988.226	12.000.331	21,92%
Índia	19.311.663	14.544	1.105.677	5,73%
Brasil	11.355.220	0	145.043	1,28%
Indonésia	9.885.081	0	362.070	3,66%
Rússia	8.948.132	0	320.088	3,58%
Alemanha	8.286.827	4.876.027	3.143.700	37,94%
Reino Unido	7.994.284	2.620.394	2.513.856	31,45%
Japão	7.146.514	6.642.428	405.834	5,68%
Canadá	6.696.763	207.354	1.423.139	21,25%

Source: WWF (2019).

Plastic is mainly used as packaging, around 36% (Figura 47). Until the 1960s, packaging was made of glass, wood or paper, materials that were gradually replaced by plastic almost entirely by the end of the 20th century.

Figura 47 - Plastic use as a percentage by type of product. Source: Gasine, 2022.

Jambeck et al (2015), estimated the annual amounts of poorly managed waste with three scenarios for what could end up in the oceans in 2025, arriving at a minimum of 10.5 million tons and a maximum of 28 million tons polluting the oceans each year.

TABELA 4 - Annual and cumulative quantities (million metric tons (MMT)) of poorly managed plastic waste and plastic marine debris (assuming three different conversion rates) for 2010-2025.

Year	Poorly managed plastic waste	25% marine debris	15% marine debris	40% marine debris
2010	31,9	4,8	8	12,7
2015	36,5	5,5	9,1	14,6
2020	41,3	6,2	10,3	16,5
2025	69,9	10,5	17,5	28
Cumulative	618,7	92,8	154,7	247,5

Source: Jambeck et al (2015).

Asian countries produce more than half (51%) of the world's plastics (Figura 48), while Latin American countries produce the least (4%).

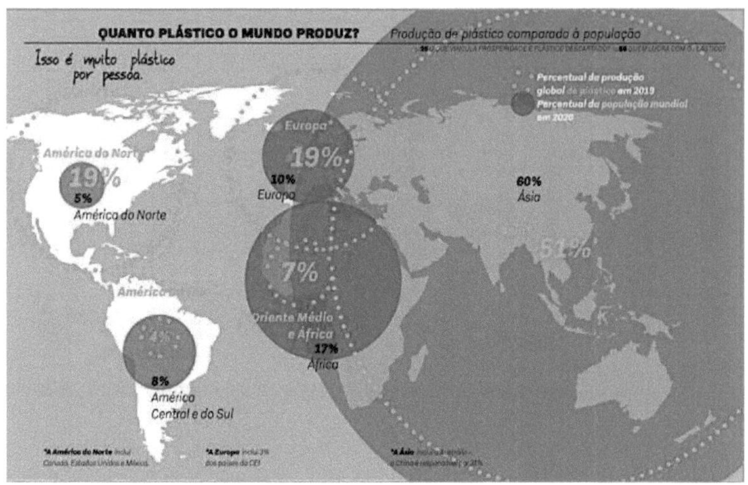

Figura 48 - Plastics production in 2019 by continent. Source: Gasine (2022).

The patch of plastic floating between the coasts of California and Hawaii is larger than the city of São Paulo; it was formed from waste dragged from the continents and discards from ships, carried by sea currents to a point of convergence (SOUZA, 2018).

Figura 49 - Plastics floating between the coasts of California and Hawaii. Source: Souza (2018)

Marine animals suffer the most from plastic pollution. Many of them swallow plastic by mistake, thinking it's food, which leads to their death (Figura 50).

Figura 50 - Turtles confuse plastic with the living water they feed on. Source: Souza (2018).

Seabirds that feed by flying over or swimming in the seas, catching small fish and crustaceans, such as the frigatebird (*Fregata magnificens*) and the albatross (Figura 51), confuse plastic waste with the animals they feed on and, unable to digest it, end up dying.

Figura 51 - Albatross killed by ingesting plastic waste. Source: Pedro (2016).

Whales swallow plastic by mistake and can no longer feed, dying of hunger and dehydration (Figura 52).

Figura 52 - Whale found dead with 40 kilos of plastic in its stomach. Source: G1-Nature (2019).

A sperm whale has been found off the coast of Murcia, in south-eastern Spain. The Centro de Recuperación de Fauna Silvestre El Valle pointed to gastric shock as the cause of death. Inside the whale were found bags, nets, ropes and other plastic products. The inner walls of the whale's abdomen were inflamed due to an infection caused by bacteria or fungi, probably as a result of ingesting the plastic and the difficulty in expelling it from the body (HYPENESS, 2018).

Figura 53 - Source: Hypeness (2018).

Recent issues linked to the by-product called micro or nanoplastics have attracted the attention of the scientific community. Microplastics are classified as primary (Figura 54) and secondary (Figura 55).

Figura 54 - Pellets identified at a collection point on Praia do Galeão, in Guanabara Bay. Source: Neves et al (2022).

Primary microplastics are particles with original characteristics produced for specific use in a tiny size, such as the microspheres found in personal care products, skin cleansers, plastic pellets for industry and fibers for synthetic fabrics. Secondary microplastics result from the degradation and fragmentation of larger products by mechanical stress and photodegradation (NEVES et al (2022).

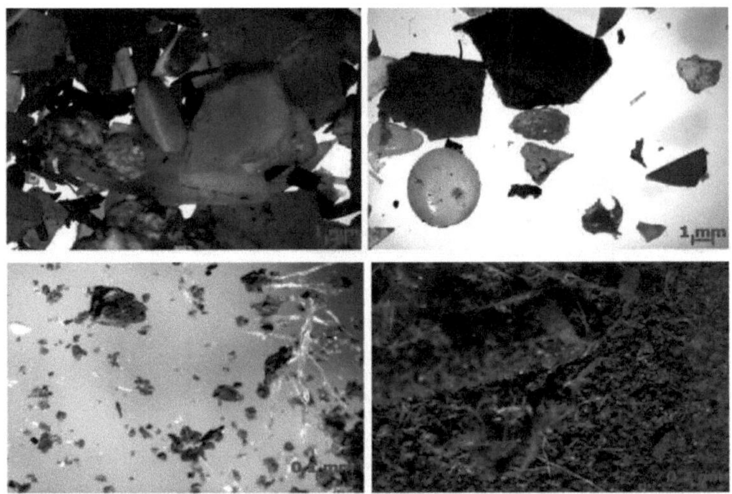

Figura 55 - Secondary plastic microparticles identified at a collection point at Praia do Galeão, in Guanabara Bay. Source: Neves et al (2022).

Neves et al (2022) report some of the impacts of microplastics on biota in the following text:

> "The progression of organic compounds through the trophic network can result in their biomagnification (Kelly et al., 2007). Plastic microparticles carrying pollutants can reach amphipods (Chua et al., 2014), polychaetes (Besseling et al., 2013), mussels (Avio et al., 2015) and fish (Oliveira et al., 2013). Other authors have reported an increase in the bioavailability of pollutants following the assimilation of microplastics through physiological responses such as inflammation (Von Moos et al., 2012) and a stimulated immune response (Browne et al., 2008). The results can be reduced feeding capacity (Besseling et al., 2013), decreased energy reserves and reproductive capacity (Sussarellu et al., 2016) and mortality (Browne et al., 2013). The exact environmental impacts of trophic transfer remain unknown (NEVES et al, 2022)."

Microplastics can be transported by wind and rain (Figura 56).

Figura 56 - Microplastics are carried by the wind and rainwater, contaminating the soil, vegetation and bodies of water where they are deposited. Source: Neves et al, 2022.

According to Brahney et al (2020), no place is safe from plastic pollution and showed that isolated areas of the United States such as national parks and national wilderness areas have accumulated microplastic particles after being carried by wind and rain. The authors estimated that more than 1,000 metric tons a year fall into protected areas in the south and mid-west of the US, the majority of which are synthetic microfibers used in the manufacture of clothing. They also showed that urban centers and soil or water resuspension are the main sources of wet deposited plastics and that plastics deposited in dry conditions were smaller in size. Deposition rates averaged 132 plastics per square meter per day, which is equivalent to more than 1,000 metric tons of annual plastic deposition on protected lands in the western US.

6.8.4 Medicines

Drugs have been detected in sewage treatment plants, surface water, groundwater and drinking water in countries such as Germany, Brazil, Canada, the United States, the Netherlands, England and Italy. People and animals treated with medicines excrete some of the drugs into the domestic sewage system or into the soil. Drug residues are carried into the aquatic environment, including parts not retained in treatment plants (BILA and DEZOTTI, 2016). A Figura 57 shows the possible routes of pharmaceutical waste to the population's treated water supply.

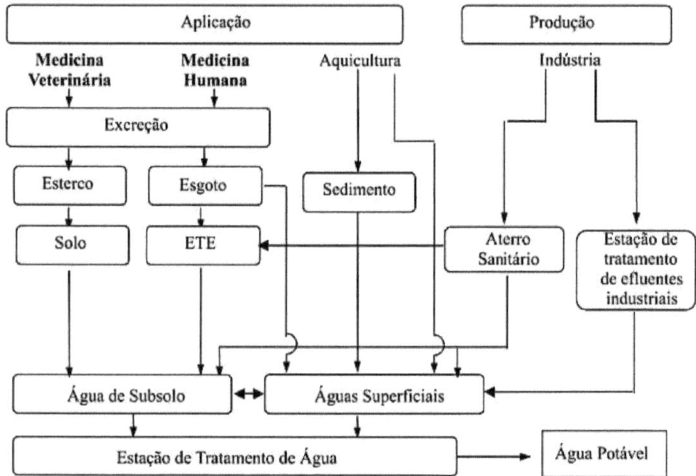

Figura 57 - Possible routes of the drugs in the environment to the treated water supplied to the population. Source: Bila and Dezotti (2016).

According to Carvalho (2023), Brazilian cities could generate up to 5,800 tons of pharmaceutical waste between 2014 and 2018. These could include antibiotics, hormones, anesthetics, antilipemics, X-ray contrast media, anti-inflammatories, among others, which have already been detected in domestic sewage, surface water and groundwater around the world (BILA and DEZOTTI, 2016).

These substances are invisible in the environment and not much is known about their effects, except for some research published in recent years. These studies have revealed changes in the biology of living beings (mainly aquatic species), causing physiological and genetic alterations through genotoxicity, gene and chromosome mutations, the development of greater resistance to antibiotics by pathogenic bacteria, endocrine disruption, causing gender alterations, damaging the reproduction and health of animals and plants and inducing diseases such as cancer.

6.8.5 Organic waste

Organic waste is biodegradable and, when improperly disposed of in the environment, generates leachate (Figura 59) which contaminates surface water and groundwater, forms an environment conducive to the proliferation of disease vectors, produces gases such as methane and carbon dioxide that are released into the atmosphere (CASTRO, 2021), as well as becoming food for certain species that can cause other problems (Figura 58).

Figura 58 - Environmental impacts caused by a landfill. Source: IFSUL (2024)

Figura 59 - Manure produced in dumps contaminates the soil, groundwater and surface water.

Disposal near airfields, for example, causes a concentration of vultures and herons (Figura 60) which cause incidents with airplanes during take-offs and landings.

Figura 60 - The dump near Marechal Rondon Airport in Várzea Grande was 20 km from the airport and the birds posed a risk to aviation. Source: G1 (2012).

6.8.6 Metals

The contamination of water and soil by heavy metals is one of the main environmental problems of the 21st century. Some metals are essential in small concentrations for the health and growth of organisms, such as sodium, potassium, calcium, iron, zinc, copper, nickel and magnesium. However, they are very toxic chemical elements in high concentrations, even chromium, zinc, iron, cobalt, manganese and nickel, which are necessary for living beings, affect them negatively in high concentrations. Heavy metals such as mercury, cadmium, chromium and lead are harmful to life. Others such as arsenic, aluminum, titanium, tin and tungsten also negatively affect life above certain concentrations in the environment (Kawai et al, 2024; SOUZA et al,2018).

The main sources of heavy metals are listed in Tabela 5.

TABELA 5 - Main sources of heavy metals

Metal	Main Sources
Lead	- automotive battery industry, semi-finished sheet metal, metal pipes, cable sheating, gasoline additives, ammunition. - automotive battery scrap recycling industry for the reuse of lead.
Cadmium	- smelting and refining of metals such as zinc, lead and copper - cadmium derivatives are used in pigments and paints, batteries, electroplating processes, soldering, accumulators, PVC stabilizers, nuclear reactors.
Mercury	- mining and the use of by-products in industry and agriculture - salt electrolysis cells for chlorine production.
Chrome	- leather tanning, electroplating.
Zinc	- metallurgy (smelting and refining), lead recycling industries.

Source: Kawai et al (2024).

The main impacts of heavy metals in the environment are soil, water and air pollution (SOUZA et al, 2018), contaminating living organisms and, due to their bioaccumulative effect, contaminating trophic chains (Tabela 6).

According to Kawai et al (2024), Brazil produces 2.9 million tons of hazardous industrial waste every year, of which only 600,000 tons receive adequate treatment; the remaining 78% is improperly deposited in dumps.

In addition, many products made up of metals such as parts, equipment, vehicles, packaging, among others, are disposed of improperly after use, or stored out in the open for later recycling, contaminating the soil, water and air.

TABELA 6 - Effects of heavy metals on human health and the environment.

Effect	Metal		
	Lead (Pb)	**Mercury (Hg)**	**Cadmium (Cd)**
In health	It causes changes in the blood and urine, leading to serious illnesses and, in some cases, total and irreversible disability. Causes respiratory problems. Causes renal and neurological changes. The main changes are in children's brain development, which can lead to idiocy. Although less aggressive in water than in air, deposited in bones, muscles, nerves and kidneys, it causes agitation, epilepsy, tremors, loss of intellectual capacity and anemia.	It affects the central nervous system, causing lesions in the cortex and granular layer of the brain. Changes in the organs of the cardiovascular system. It accumulates in the nervous system, mainly in the brain, spinal cord and kidneys. It causes loss of coordination of movements, difficulty in speaking, eating and hearing, as well as atrophy and kidney, urogenital and endocrine damage.	It causes changes in the central nervous system and the respiratory system. It compromises bones and kidneys. Causes pulmonary edema, lung cancer and irritation of the respiratory tract. Similarly to mercury, it affects the nervous system and kidneys. It causes loss of smell, the formation of a yellow ring on the neck of the teeth, a reduction in the production of red blood cells and the removal of calcium from the bones.
In the environment	It pollutes the soil, water and air and thus contaminates living organisms due to its bioaccumulative effect throughout the food chain (trophic).	It is absorbed by living organisms and accumulates continuously throughout life. Through contamination of water or soil, it easily enters the food chain, posing a danger to humans who feed on fish or birds from these areas.	Contaminates soil, air, water and groundwater. It bioaccumulates throughout the food chain (trophic), causing poisoning in humans when they eat fish contaminated with cadmium.

Source: Kawai et al (2024).

Illegal gold mining, in addition to destroying river beds and banks, is a source of mercury contamination in the water (Figura 61), which enters the food chain and contaminates fish and other aquatic animals.

Figura 61 -Illegal gold mining .

Metal mining produces acid mine drainage, which is the formation and movement of highly acidic water, rich in heavy metals, generated mainly from mining waste (BLOG ERA DA ÁGUA, 2022).

Figura 62 - Acid mine drainage pollution. Source: Water Age Blog (2022).

In addition to the metallurgical industry, mining, dumps, irregular deposits of metallic material (Figura 63) and electronic waste (Figura 64) cause intense pollution.

Figura 63 - Irregular deposit of scrap metal contaminating the environment with heavy metals.

Figura 64 - Irregular disposal of electronic waste.

One of the biggest issues with metals is that the amounts present in the environment tend to be increasing and remain in trophic chains, as they cannot be metabolized and accumulate in ecosystems (SOUZA et al, 2018).

In addition, there have been environmental disasters such as the collapse of the Fundão Dam, an iron ore processing tailings dam (Mariana-MG) in November 2015, which polluted the Doce River (Figura 65) over a length of more than 600 km and reached the sea, which was also polluted by the mud

116

laden with heavy metals from the mining operations promoted by Samarco Mineração S.A (MPF, 2024).

Figura 65 - The Mariana disaster led to the degradation of the Doce River and many impacts remain to this day. Source: Alencar (2018).

6.8.7 Radioactive Elements

Radioactive elements are those with an atomic number greater than or equal to 84 and radioactivity is represented by a trefoil as in Figura 66.

They are used for the following purposes:

> Energy production

> Agriculture

> Industrial

> Medicine

> Environmental research

> Nuclear weapons

Figura 66 - Trifolium is used as a symbol of radioactivity.

Common applications of radioactive elements:
- ➤ Uranium-235: used in nuclear reactors;
- ➤ Technetium-99: used in imaging studies of the brain, lungs and liver;
- ➤ Iodine-131: used in the post-surgical treatment of thyroid carcinoma;
- ➤ Plutonium-239: used to make nuclear weapons and as fuel for nuclear reactors;
- ➤ Samarium-153: used in the treatment of bone cancer.

Types of radiation are as follows:
- ➤ Cosmic radiation
- ➤ X-rays
- ➤ Gamma Radiation (γ)
- ➤ Alpha radiation (α)
- ➤ Beta radiation (β)
- ➤ Neutron radiation (n)
- ➤ UV radiation (natural)

Uncontrolled exposure to radioactivity is harmful to living beings and the most common radioactive pollutants are the following: atomic waste (Figura 67), waste from atomic explosions, discarded equipment and nuclear reactors.

Figura 67 - Representation of nuclear power plant waste.

6.9 Climate change caused by civilization

The Earth's climate has remained stable since the last glaciation. Plants and animals have adapted to their ecosystems and any change puts many living beings at risk of migration, maladaptation and extinction. Environmental catastrophes caused by climate change such as extreme temperatures and uncontrollable fires, an increase in the number and intensity of hurricanes and tornadoes, floods, blizzards, melting glaciers and ocean ice sheets and rising sea levels cause problems for animal and plant life, unbalancing and destroying habitats, causing extreme hunger and thirst and even the death of many individuals and populations (Figura 68).

The main causes of climate change related to human activity are the burning of fossil fuels, inadequate land use, fires and forest degradation (CASTILHO, 2024).

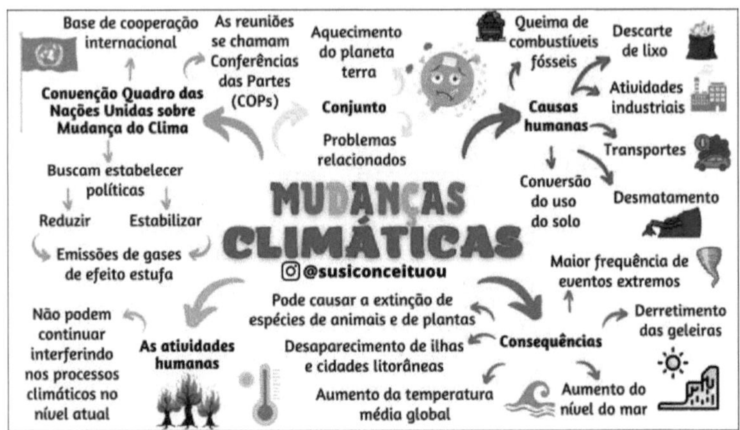

Figura 68 - Climate change, causes and consequences. Source: Study Maps (2023).

The risks of climate change for species in WWF conservation priority areas are listed below, according to Warren et al (2018):

"FUTURE OF EXTREMES - Even with the emissions reductions promised under the Paris Agreement, temperatures that were once extreme are set to become the "new normal" in all priority areas.

GLOBAL WARMING - Temperature increases have been observed in all seasons in WWF priority areas over the last 50 years.

2°C INCREASE IN TEMPERATURE - If the increase in global temperature is limited to 2°C, less than 25% of species in priority areas will be at risk of local extinction.

50% LOSS OF SPECIES - If global temperatures rise by 4.5°C, almost 50% of species in priority areas will be at risk of local extinction.

CHAIN EFFECTS - More than 50% of plant species could become extinct in areas with a large increase in temperature, which would have detrimental effects on many other species.

DISPERSION CAN HELP - With a 2°C increase in global temperature, the risk of local extinction drops to around 25% without dispersal and to 20% if species can move freely.

PRESERVATION OF HABITATS - With a global temperature increase of 2°C, 56% of priority areas will remain climatically suitable for species. With an increase of 4.5°C, this figure will be just 18%."

More than a third of Amazonian species could become extinct by 2080 with a 2°C increase in the average global temperature; if the animals disperse to areas where they can adapt better, the scenario is slightly better (Tabela 7), but it is still very bad in many cases (WARREN et al, 2018).

TABELA 7 - Percentage of species in the Amazon that are expected to be at risk of local extinction by 2080 in three different global warming scenarios, modeled both with dispersal and without dispersal

| | Cenário das mudanças climáticas globais | | | | | |
| | 2°C | | 3.2°C | | 4.5°C | |
Grupo de espécies	Sem dispersão	Com dispersão	Sem dispersão	Com dispersão	Sem dispersão	Com dispersão
Plantas	43	43	59	59	69	69
Pássaros	37	+	51	+	64	13
Mamíferos	36	0	50	10	63	30
Anfíbios	47	47	62	62	74	74
Répteis	35	35	48	48	62	62

Source: Warren et al (2018). ('+' indicates a possible increase in richness caused by the colonization of other species).

The causes of current climate change have mainly been attributed to human activities. However, changes in the earth's climate have occurred throughout the planet's existence. Periods of volcanism, changes in solar activity, meteor impacts and the activity of living organisms over long periods such as stromatolites have transformed both the climate and the composition of the atmosphere with extreme consequences for living beings.

6.10 Natural disasters

These are natural phenomena that cause great destruction, loss of life, alterations to the earth's surface and destruction of ecosystems, such as those caused by extreme weather events, volcanic eruptions, earthquakes and tsunamis.

6.10.1 Earthquakes and tsunamis

The movement of tectonic plates generates earthquakes that cause abrupt movements of the earth's crust and often occur near or under the ocean, generating tsunamis that cause great destruction, affecting everything they pass through.

6.10.2 Volcanism

Intense volcanism in certain periods of the earth's existence has caused radical changes in the earth's climate and atmosphere. Two events show how impactful volcanism has been in the past:

> ➢ During the Cryogenian period, between 790 and 635 million years ago, the Earth was completely covered in ice and the thawing of the Earth is attributed to the increase in CO_2 in the atmosphere caused by intense volcanic activity (JOEL, 2019).

> ➢ According to Hall (2018), there is evidence that volcanoes were the real culprits for the mass death of several species before the meteor impact on the Yucatan Peninsula in Mexico 66 million years ago; the author cites that: "volcanic activity struck the first blow, weakening the climate so much that a meteor - the most impactful blow - was able to cause disaster for *Tyrannosaurus rex* and its late cretaceous relatives."

Another aspect of volcanism is its direct impact on the organisms that live around eruptions, caused by pyroclastic flows and lava spills, which can extinguish endemic species or reduce the density of populations. An example of recent destruction occurred in 2021 with the eruption of the Cumbre Vieja volcano, whose lava destroyed an area of 1,241 hectares, according to the Copernicus satellite monitoring system (MOSQUERA, 2021).

Figura 69 - Lava from the Cumbre Vieja volcano reaching the sea.

6.10.3 Extreme weather events

Extreme weather events (Figura 70) such as hurricanes and tornadoes, floods, extreme droughts and uncontrollable fires have caused loss of life in all environments.

Figura 70 - Extreme weather events in South America. Source: Ecoa (2023).

Hurricanes affect ecosystems in large areas, such as Katrina in Florida in 2005 (Figura 71).

Figura 71 - Hurricane Katrina - Florida, 2005. Source: NOA News, 2005.

Hurricane Katrina caused fragmentation and stress to the coral reefs it passed through, swept out to sea and buried green turtle (*Chelonia mydas*) and loggerhead turtle (*Caretta caretta)* nests and leatherback turtle hatchlings. The hurricane passed through the Everglades National Park, bringing torrential rain to an area where water levels were already high, and destroying around 90% of the *alligator* nests (*Alligator mississippiensis*); also worrying is the effect on populations of the rare Miami blue butterfly (*Hemiargus thomasi bethunebakeri*), reintroduced into the Everglades and Biscayne National Parks affected by the hurricane - the larvae and pupae of this species of butterfly are laid in the undergrowth, at risk of being swept away by the waters (GEO-RISKS, 2024).

Extreme weather events have been intensified by current climate change, but they were already occurring with less frequency and intensity before man-made changes were noticed.

Extreme droughts and high temperatures have their impacts magnified by the resulting fires, which also kill vegetation and animals (Figura 72) due to lack of food, water and hyperthermia. The water and climate crisis in the Río de la Plata Basin, due to the reduction in rainfall between 2020 and 2021, has caused environmental, social and economic impacts such as fires in the Pantanal, losses in agricultural production and a reduction in electricity generation at the dams (ECOA, 2023).

Figura 72 - Animals killed by drought in the Pantanal of Mato Grosso.

Extreme rainfall opens up gullies on the slopes, causing landslides and flooding (Figura 73), destroying vegetation and killing animals.

Figura 73 - Natural areas, pastures and crops flooded in southwestern Tocantins in 2018.

6.11 Pests and diseases

6.11.1 Pests

Pests are harmful animals that attack plants, animals and humans and can cause damage directly by harming, injuring or killing other species, or indirectly by transmitting diseases. Most pests are invasive exotic animals such as wild boar.

Approximately three quarters of the negative impacts of invasive species occur on land, especially in forests, woodlands and cultivated areas (DEUTSCHE WELLE, 2023).

Furthermore, according to Deutsche Welle (2023), "a team of 86 experts from 49 countries analyzed the global impacts of some 3,500 harmful invasive species over four years. The report estimates that these species play a central role in 60% of the recorded extinctions of plants and animals."

But native animals can also become pests due to some environmental imbalance. One example is the proliferation of capybaras due to the local

126

absence of their main predator, the jaguar. Populations of invertebrates in large numbers, such as mites, ants, hematophagous mosquitoes, grasshoppers, aphids, bedbugs, ticks, among others, can become pests of animals or plants, causing damage both in urban areas and in rural and natural areas. Birds and mammals such as starlings and hematophagous bats can also cause damage to nature.

6.11.2 Diseases

Disease is the biological alteration of a person's state of health. Diseases have different causes and can be acquired directly through contact with the etiological agent, or through the transmission of the etiological agent by other animals, such as ticks and mosquitoes. The etiological agent or pathogen can be a virus, bacterium, fungus, protozoan or helminth.

The DGAV (2024) publishes a list of wild animal diseases, as follows:

"1. Birds

- Botulism
- Newcastle disease
- West Nile fever
- Avian flu
- Other

2. Carnivores

- Echinococcosis-Hydatidosis
- Leishmaniasis
- Anger
- SARS-CoV-2
- Other

3. Lagomorphs (rabbits, hares and ocotonids)

- Cysticercosis
- Viral Hemorrhagic Disease
- Myxomatosis

- SARS-CoV-2

- Tularemia

- Other

4. Ungulates

- Brucellosis

- Cysticercosis

- Aujeszky's disease

- Chronic wasting disease in deer

- Epizootic hemorrhagic disease

- Echinococcosis-Hydatidosis

- Foot and Mouth Disease

- Bluetongue

- Rift Valley Fever

- Peste des petits ruminants

- African swine fever

- Classical Swine Fever

- Scabies

- Trichinellosis

- Tuberculosis

- Sheep and goat pox."

In Brazil, diseases such as yellow fever and bird flu have caused concern because they are attacking wild animals.

Wild animals can also be carriers of zoonoses. ACHA and SZYFRES (1986) *apud* Silva (2024) described the main conditioning factors for the spread of risk factors existing in natural outbreaks, with the possibility of establishing zoonotic processes:

"a) introduction of domestic animals and/or man into a natural outbreak;

b) translocation from an infected host to a new biotype where there are susceptible hosts;

c) modification of host dynamics or alteration of the ecological balance;

d) lack of food, which forces reservoir animals to move to other biocenoses;

e) human intervention in the modification of ecosystems;

f) positive mutations in the epidemic process of the etiological agent, facilitating its spread;

g) intervention by migratory birds and vectors".

Avian influenza, caused by the H5N1 virus, has affected birds in the Taim Ecological Station (RS), killing dozens of animals (SOUZA, 2023), such as the swan in the Figura 74.

Figura 74 - Black-necked swan killed by bird flu in Taim. Source: Souza (2023).

We already had 70% of the primates in the Atlantic Forest threatened with extinction. Now, yellow fever (Figura 75) has caused the death of primates in Brazil and could accelerate the extinction of some already threatened species (CARNEIRO, 2017).

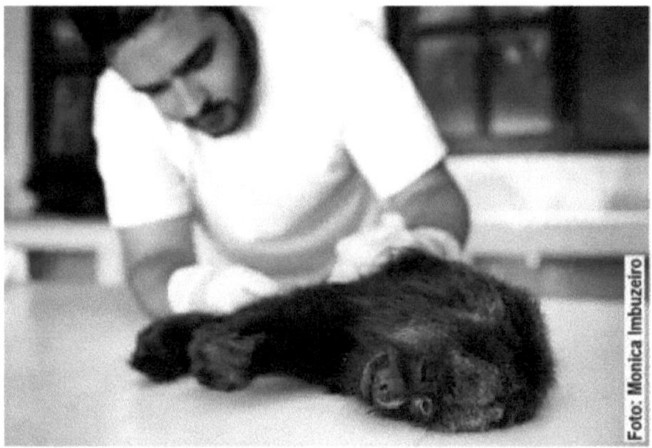

Figura 75 - Body of a howler monkey killed during the yellow fever epidemic being examined by a researcher. Source: Azevedo (2023).

6.12 Endemism, genetic diversity and reproductive rate

The survival of species and the conservation of biodiversity is all the easier the greater the biodiversity of populations and the greater their genetic variation. Therefore, endemism and low genetic diversity go against the grain of maintaining biodiversity.

Endemism is the occurrence of a species exclusively in a small geographical area, usually in a single location or ecosystem. Endemic species have adapted to the specific environmental conditions of the ecosystems where they live, making them vulnerable to environmental changes such as climate change, habitat destruction and the introduction of invasive species. Consequently, the greater the isolation and the more specific the environmental conditions of a given location, the greater the chances of endemism.

Greater genetic diversity makes it easier for species to adapt and survive over time. The greater the genetic diversity of a population, the more likely it is to be able to resist disease and adapt to changes in the environment. Small, isolated populations, such as those of endemic species, generally have reduced genetic diversity. Low genetic diversity limits the ability to adapt to environmental changes. The occurrence of a serious disease epidemic, or a strong change in climate, can mean that a genetically poor population cannot survive the new conditions, making it susceptible to extinction.

Small, isolated populations can have a high degree of inbreeding, increasing the likelihood of genetic defects and health risks and reducing their adaptability.

Another risk factor is the species' reproductive rate; when it is high, population growth generates a large number of individuals who can undergo beneficial mutations that increase the species' chances of adaptation and survival. Low reproductive rates represent a risk of population reduction due to disease and maladaptation, which could lead to extinction.

7 Extinctions caused by man

The extinction of species is part of natural processes: more adaptable and resistant species survive environmental changes, while those with less adaptability succumb. Excessive hunting, habitat destruction, pollution and climate change have accelerated extinction rates across the globe. Some examples of animals extinguished by man are listed below.

Initially it was hunting, associated with climate change at the end of the last glaciation 10,000 years ago. More recently, several species hunted intensively after the invention of firearms have become extinct. And the intense and rapid modification of the earth's environment caused by human civilization since the industrial revolution, starting in the second half of the 18th century, has prevented many animal species from adapting to the new conditions and we have been experiencing a drastic reduction in wild species populations since the second half of the 20th century.

7.1 Woolly mammoth (*Mammuthus primigenius*)

The extinction of the woolly mammoth (Figura 76) occurred around 10,000 years ago. They lived in the Pleistocene in vast areas of the northern hemisphere. It became extinct due to excessive hunting by humans, associated with drastic climate changes at the end of the last glaciation.

Figura 76 - Woolly mammoth (Mammuthus primigenius). Source: Galileo Magazine.

7.2 Traveling pigeon (*Ectopistes migratorius*)

The traveling pigeon (Figura 77) was an extremely numerous bird in North America, with flocks that darkened the sky. Uncontrolled hunting for sport and food and the destruction of its habitat led to its extinction in the wild around 1900 and the last known individual, named Martha, died in 1914 at the Cincinnati Zoo.

Figura 77 - Traveling pigeon (*Ectopistes migratorius*); this taxidermied specimen is part of the collection of extinction expert Errol Fuller. Source: National Geographic, 2028.

7.3 Java Tiger (*Panthera tigris sondaica*) and Bali Tiger (*Panthera tigris balica*)

Java tigers and Bali tigers were subspecies of tigers that inhabited the Indonesian islands of Java and Bali. The destruction of their habitats by agricultural expansion and illegal hunting led to the suppression of their populations. The Bali tiger probably became extinct in the 1970s and the Java tiger in the 1980s.

7.4 Western black rhino (*Diceros bicornis longipes*)

The western black rhino (Figura 78) was a subspecies of West African black rhino and was last found in Cameroon. Illegal hunting has caused its population to decline. In 2011, the western black rhino was declared extinct by the IUCN.

Figura 78 - Western black rhino (*Diceros bicornis longipes*).

7.5 Chinstrap dolphin (*Lipotes vexillifer*)

The Chinese fairy dolphin, or Yang-Tsé dolphin, also known as baiji and Qi Qi (Figura 79), was a species of dolphin that occurred by the thousands along 1,700 kilometers of the middle and lower Yangtze in China.

Figura 79 - Qi Qi, baiji who lived more than 15 years in captivity. Source: O Globo (11/10/2016).

Pressure from the human population, the construction of the Three Gorges Hydroelectric Power Station, as well as other smaller dams, has led to habitat loss, and together with water pollution and over-hunting, has drastically reduced the baiji population in the river. It was last sighted in 2004 and is thought to be extinct.

7.6 Other animals made extinct by man

Other animals that have become extinct as a result of human activity are listed below:

> Blue Antelope (*Hippotragus leucophaeus*) - Extinct around the year 1800.
> Lesser Bilby (*Macrotis leucura*) - Extinct in the 1950s.
> Schomburgk deer (*Rucervus Schomburgki*) - Extinct in 1938.
> Steller's dugong (*Hydrodamalis gigas*) - Extinct around 1768.
> Norfolk Kaka (*Nestor productus*) - Extinct in the wild in the 19th century.
> Katydid (*Neduba extincta*) - An insect from the Orthoptera order that became extinct in 1996.
> Stephen's Island lark (*Xenicus lyalli*) - Extinct in 1894.
> Pyrenean ibex (*Capra pyrenaica pyrenaica*) - Extinct in 2000.

➢ Woodcock (*Xenicus longipes*) - Extinct in 1972.

➢ Tarpon (*Equus ferus ferus*) - A species of wild horse that became extinct in 1909.

➢ Atlas lion (*Panthera leo leo*) - Extinct from the wild in the 1940s.

8 RED BOOK

The International Union for Conservation of Nature (IUCN) is a civil organization based in Gland, Switzerland, dedicated to nature conservation, founded in 1948.

The IUCN is a Union of members made up of governmental organizations and civil society. With more than 1,400 member organizations and the contribution of around 15,000 experts, the IUCN is the global authority on the state of the natural world and the measures needed to protect it.

The IUCN's work focuses on a series of initiatives, including the creation of natural parks for environmental preservation and the drawing up of the **Red List of Threatened Species,** created in 1964. The list is a detailed inventory of the conservation status of species of plants, animals, fungi and protists, assessing the risk of extinction of thousands of species and subspecies, relevant to all species and in all regions of the world, with the aim of informing about the need for conservation measures, helping the international community to try to reduce extinctions. The lists also help to create conservation and management policies and plans for threatened species.

8.1 Animal species assessments in Brazil

Brazil published the first national list of endangered animal and plant species in 1968, drawn up by the now-defunct Brazilian Forestry Development Institute (IBDF, 1968). Below is a list of the normative acts that have instituted and altered lists in the country since then:

- IBDF Ordinance No. 303, of May 29, 1968 - Establishes the official Brazilian list of animal and plant species threatened with extinction in the country.
- IBDF Ordinance No. 093/80-P of February 5, 1980.

- IBAMA Ordinance No. 1522, of December 19, 1989 - Recognizes the Official List of Brazilian Fauna Species Threatened with Extinction.
- MMA Normative Instruction No. 03, of May 27, 2003 - Official List of Endangered Brazilian Fauna Species (considering only the following groups of animals: amphibians, birds, terrestrial invertebrates, mammals and reptiles);
- MMA Normative Instruction No. 05, of May 21, 2004 - Official List of Aquatic Invertebrate and Fish Species Threatened with Extinction and Overexploited or Threatened with Overexploitation;
- MMA Normative Instruction No. 52, of November 8, 2005 - Amends Annexes I and II of MMA Normative Instruction No. 05, of May 21, 2004
- IBAMA Ordinance No. 444, of December 17, 2014 - Endangered fauna;
- IBAMA Ordinance No. 445, of December 17, 2014 - Threatened fish and aquatic invertebrates;
- MMA Ordinance No. 98, of April 28, 2015.
- MMA Ordinance No. 163, of June 8, 2015.
- MMA Ordinance No. 148, of June 7, 2022 (valid).
- Ordinance GM/MMA No. 300, of December 13, 2022 (revoked).
- MMA Ordinance No. 354, of January 27, 2023 (valid).

8.2 Red Book of Threatened Fauna Species

In 2008, the first edition of the "Red Book of Endangered Brazilian Fauna" was published. Its launch coincided with the 40th anniversary of the publication of the first "Official List of Endangered Brazilian Fauna Species" (MMA, 2008).

The Red Book went further, informing us of the methodology used to process the assessments, what the legal basis is, which experts take part in the assessment process and who the validators are, providing detailed information on the animals listed, such as taxonomy, extinction risk category, geographical distribution, natural history and ethology, population, threats, conservation actions, presence in protected areas and existing and necessary research. In addition to describing the geographical distribution of the species, maps are presented, based on records of occurrence or a polygon representing the area in which the species can potentially be found (ICMBio, 2018b). The Red Lists and Red Books of Endangered Species, as well as the National Action Plans (PANs), are drawn up to guide actions and remove species from the red lists, helping to draw up and update the National Environmental Policy.

The latest edition of the Red Book of Endangered Brazilian Fauna was published in 2018 (Figura 80) and was divided into 7 volumes and the red list was updated in 2022.

Figura 80 - Red Book of Endangered Brazilian Fauna. Source: ICMBio (2018).

8.3 Risk of extinction

Understanding the conservation status of biodiversity is the starting point for planning the measures that should be taken to reduce the risk of species extinction (SOUZA et al, 2018). The risk of extinction of all vertebrates and selectively of some invertebrates that occur in Brazil is assessed using the method of categories and criteria of the International Union for Conservation of Nature - IUCN.

The assessment of the risk of extinction is an institutionalized, regular and continuous process, with standardized, standardized and documented procedures in all its stages (SOUZA et al, 2018).

8.4 Assessment of the Conservation Status of Brazilian Fauna

According to Souza et al (2018), the conservation status of Brazilian fauna is assessed as follows:
  - ➢ Evaluation of taxonomic groups as a regular and continuous process, with re-evaluation cycles every 5 years;
  - ➢ Use of the method of categories and criteria of the International Union for Conservation of Nature - IUCN to assess the risk of extinction of species.
  - ➢ The IUCN method is adopted for its comprehensive, objective and scientifically rigorous approach to any organism except microorganisms.

It is a model adopted worldwide for drawing up national lists of threatened species.

The guidelines for the evaluation process are regulated by ICMBio Normative Instruction No. 34 of 2013 and are listed below:
  - ➢ Vertebrates - all species with confirmed occurrence in Brazil are evaluated;
  - ➢ Invertebrates - a selection of taxonomic groups is made, preferably those considered to be indicators of environmental quality, important for environmental services or that are used commercially;

> A permanent network of experts has been set up in partnership with the IUCN, research institutions, scientific societies and non-governmental organizations with recognized work in biodiversity conservation;

> Conservation assessments and recommendations are based on the best available data and information, derived from the most recent scientific studies and the field experience of experts from the scientific community.

> The diagnosis of the situation of each species is made in collaboration with the researchers linked to these institutions, who actively participate in the process.

> The ICMBio team of technicians involved in carrying out the assessment process is continuously trained in the IUCN categories and criteria method;

> All stages of the process are documented.

Steps in the evaluation process are carried out according to the flow in Figura 81.

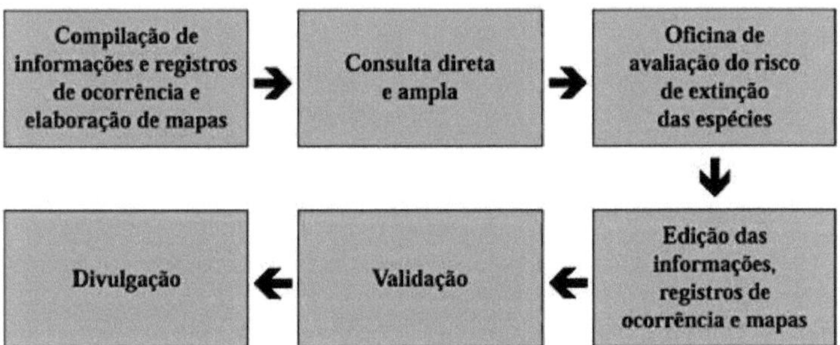

Figura 81 - Stages in the process of assessing the conservation status of Brazilian fauna. Source: SOUZA et al (2018).

The IUCN standards contained in the document "IUCN Red List Categories and Criteria, Version 3.1, Second Edition" (IUCN, 2012) and the guide "Guidelines for using the IUCN Red List Categories and Criteria, Version 15.1" (IUCN 2022) are used to apply the wildlife assessment method.

8.5 Extinction risk categories

The risk classification is carried out by combining the following information:

> ➢ Population size and information on fragmentation, fluctuations or past and/or projected decline;
> ➢ Extent of geographical distribution, area of occupation and information on fragmentation, decline or fluctuations;
> ➢ Threats affecting the species;
> ➢ Existing conservation measures.

Species considered Critically Endangered (CR), Endangered (EN) or Vulnerable (VU) are those that are considered to be under real threat of extinction; they are species that require conservation action in the immediate future. The risk categories are described according to the IUCN method used in the Process for Assessing the Conservation Status of Brazilian Fauna (IN ICMBio No. 34, 2013).

The classification of the conservation status of animal or plant species for drawing up the red list of Brazilian fauna is carried out using the dichotomous key scheme published by the IUCN in 2001 (Figura 82).

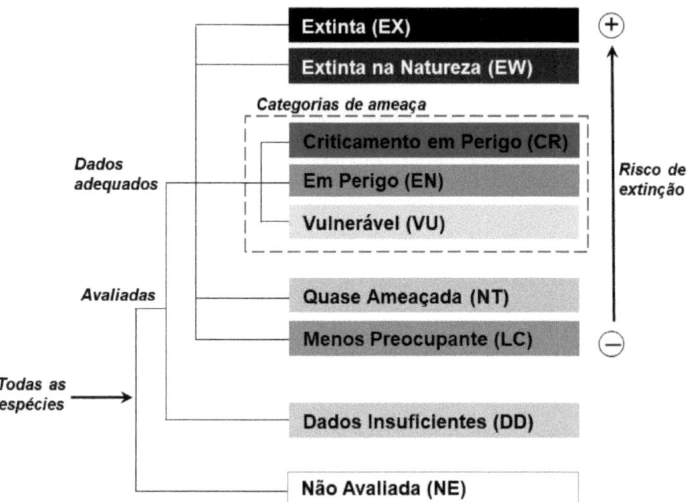

Figura 82 - Structure of the species categories evaluated according to the IUCN (2001).

8.5.1 Meaning of the IUCN Categories for Species Assessment

The taxa for which there is no longer a record in the wild are assessed as Extinct or Extinct in the Wild, according to one of the two situations below and the other categories are as described below (ICMBio, 2018):

> "**EXTINCT (EX)** - a taxon is Extinct when there is no doubt that the last individual has died. This only occurs when, after exhaustive inventories in its known and/or expected habitat at appropriate times (diurnal, seasonal, annual), throughout its historical distribution area, no individual is recorded. Surveys should be carried out on a time scale appropriate to the life cycle and life form of the taxon.
>
> **EXTINCT IN THE NATURE (EW)** - a taxon is considered Extinct in the Wild when it is known to exist only in cultivation, captivity or in populations inserted into the wild, in areas completely different from its original area of occurrence.

THREATENED: The IUCN, according to version 3.1 (IUCN, 2001), distinguishes three levels of threat for species, as follows:

- **CRITICALLY ENDANGERED (CR)** - a taxon is considered Critically Endangered when it is at extremely high risk of extinction in the wild in the immediate future.
- **ENDANGERED (EN)** - a taxon that is not Critically Endangered, but is at very high risk of extinction in the wild in the near future.
- **VULNERABLE (VU)** - a taxon that does not fall into the Critically Endangered or Endangered categories, but is at high risk of extinction in the wild in the medium term.

These three subcategories of endangered species were used to inform the decision-making of the experts who took part in the process of revising the list. The species that are at risk of extinction in Brazil, whether extreme, very high or high, are all considered "Threatened" under the current Brazilian legal framework.

In addition to these, the IUCN also uses the categories called Near Threatened and Least Concern (which we prefer to call Non-Threatened) when the available data set is considered adequate for what is required in the assessment, as well as the Data Deficient category, where, as the name of the category indicates, the current information on the taxon does not meet the minimum requirements necessary for it to fall into any of the categories defined for an adequate level of knowledge. See the meaning of these categories:

a) **Near Threatened (NT)**: A taxon is Near Threatened when it does not meet the criterion for threatened - vulnerable - but is very close to it, so that if it is not protected, it will quickly become threatened.

b) **Non-threatened (LC)**: This means a species that has been assessed and for which the existing information does not justify its inclusion in one of the risk categories, according to the criteria adopted. This category includes taxa with a wide distribution and high abundance, which is why they are considered to be of Least Concern or, as has been adopted in Brazil, Not Threatened.

c) **Data Deficient (DD)**: A taxon is considered so when the existing data on it does not allow us to know whether or not it is threatened. Species in this category require more research in order to reach a firm conclusion about their conservation status."

The species that are indicated for inclusion on the Official National List of Endangered Brazilian Fauna Species are those categorized as Extinct in the Wild (EW), Critically Endangered (CR), Endangered (EN) and Vulnerable (VU).

Although species categorized as Insufficient Data (DD) are not on the list of endangered species, they may be at risk of extinction, but current knowledge does not allow such a statement, requiring further studies and investigations for a more conclusive analysis.

The criteria for risk classification are as follows:

A. Population reduction (past, present and/or projected) (Tabela 8);

B. Restricted geographical distribution and showing fragmentation, decline or fluctuations (Tabela 9);

C. Small population with fragmentation, decline or fluctuations (Tabela 10);

D. Very small population or very restricted distribution (Tabela 11);

E. Quantitative extinction risk analysis (e.g. PVA - Population Viability Analysis) (Tabela 12).

TABELA 8 - Population reduction (past, present and/or projected)

A. Redução da População (Declínio medido ao longo de 10 anos ou 3 gerações, o que for mais longo)			
	Criticamente Em Perigo	Em Perigo	Vulnerável
A1	$\geq 90\%$	$\geq 70\%$	$\geq 50\%$
A2, A3 e A4	$\geq 80\%$	$\geq 50\%$	$\geq 30\%$

A1 Redução da população observada, estimada, inferida ou suspeitada de ter ocorrido no passado, sendo as causas da redução claramente reversíveis E compreendidas E tenham cessado.
A2 Redução da população observada, estimada, inferida ou suspeitada de ter ocorrido no passado, sendo que as causas da redução podem não ter cessado OU não ser compreendidas OU não ser reversíveis.
A3 Redução da população projetada ou suspeitada de ocorrer no futuro (até um máximo de 100 anos).
A4 Redução da população observada, estimada, inferida, projetada ou suspeitada, sendo que o período de tempo deve incluir tanto o passado quanto o futuro (até um máximo de 100 anos), e as causas da redução podem não ter cessado OU não ser compreendidas OU não ser reversíveis.

baseado em um ou mais dos seguintes itens:

(a) observação direta;

(b) índice de abundância apropriado para o táxon;

(c) declínio na área de ocupação, extensão de ocorrência e/ou qualidade do habitat;

(d) níveis reais ou potenciais de exploração;

(e) efeitos de táxons introduzidos, hibridação, patógenos, poluentes, competidores ou parasitas.

Source: Souza et al, 2018.

TABELA 9 - Restricted geographical distribution and showing fragmentation, decline or fluctuations

B. Distribuição geográfica restrita e apresentando fragmentação, declínio ou flutuações			
	Criticamente Em Perigo	Em Perigo	Vulnerável
B1 Extensão de ocorrência	< 100 km²	< 5.000 km²	< 20.000 km²
B2 Área de ocupação	< 10 km²	< 500 km²	< 2.000 km²
E pelo menos 2 dos seguintes itens:			
(a) População severamente fragmentada, OU número de localizações	= 1	≤ 5	≤ 10
(b) declínio continuado em um ou mais dos itens: (i) extensão de ocorrência; (ii) área de ocupação; (iii) área, extensão e/ou qualidade do habitat; (iv) número de localizações ou subpopulações; (v) número de indivíduos maduros.			
(c) flutuações extremas em qualquer um dos itens: (i) extensão de ocorrência; (ii) área de ocupação; (iii) número de localizações ou subpopulações; (iv) número de indivíduos maduros.			

Source: Souza et al, 2018.

TABELA 10 - Small population with fragmentation, decline or fluctuations

C. Tamanho da população pequeno e com declínio			
	Criticamente Em Perigo	Em Perigo	Vulnerável
Número de indivíduos maduros	< 250	< 2.500	< 10.000
E C1 ou C2			
C1 Um declínio continuado observado, estimado ou projetado de pelo menos (até um máximo de 100 anos no futuro):			
	25% em 3 anos ou 1 geração	20% em 5 anos ou 2 gerações	10% em 10 anos ou 3 gerações
C2 Um declínio continuado observado, estimado, projetado ou inferido E pelo menos uma das 3 condições:			
(a) (i) número de indivíduos maduros em cada subpopulação:	≤ 50	≤ 250	≤ 1.000
(ii) ou % indivíduos em uma única subpopulação	90–100%	95–100%	100%
(b) flutuações extremas no número de indivíduos maduros			

Source: Souza et al, 2018.

TABELA 11 - Very small population or very restricted distribution

D. População muito pequena ou distribuição muito restrita			
	Criticamente Em Perigo	Em Perigo	Vulnerável
D Número de indivíduos maduros	< 50	< 250	D1. < 1.000
D2 Área de ocupação restrita ou número de localizações, sob uma ameaça futura plausível de levar o táxon à condição de CR ou EX em curto prazo.	-	-	D2. Tipicamente AOO < 20 km² ou Número de localizações ≤ 5

Source: Souza et al, 2018.

TABELA 12 - Quantitative analysis of extinction risk

E. Análises quantitativas			
	Criticamente Em Perigo	Em Perigo	Vulnerável
Indicando que a probabilidade de extinção na natureza é de:	≥ 50% em 10 anos ou 3 gerações	≥ 20% em 20 anos ou 5 gerações	≥10% em 100 anos

Source: Souza et al, 2018.

In 2022, a total of 50,313 species of plants and 127,830 species of animals were thought to be recognized in Brazil, according to the IBGE (2023):

> To date, 127,830 valid animal species are known for Brazil, the vast majority of which are arthropods (around 85%, almost 94,000 species!) and chordates (around 10%). All the other species represent other groups of invertebrates. In general, except for a few phyla, the number of species for the vast majority exceeded those presented in recent estimates. Particularly noteworthy are the *Annelida* (with around 1,600 species), *Mollusca* (with almost 3,100 known valid species), Aves (almost 3,000), bony fish (around 4,400) and amphibians (just over 1,000 species)."

8.5.2 Results of extinction risk assessments for Brazilian species

Around 11% of animal species and 15% of plant species have been assessed on threatened species lists (Figura 83), showing that we are only just beginning to assess species in terms of threats and extinction risks (IBGE, 2023).

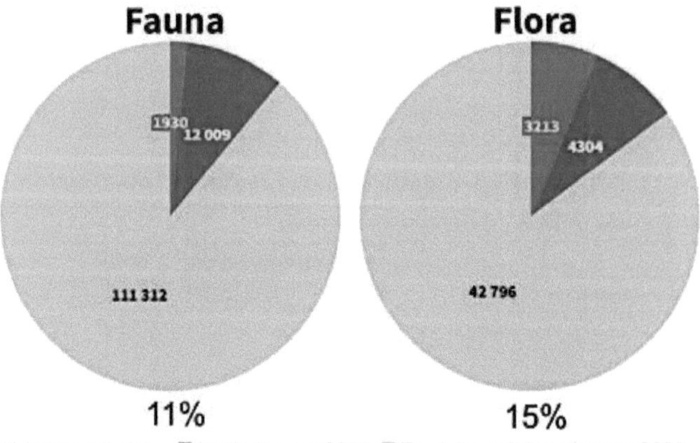

Figura 83 - Species evaluated for endangered species lists. Source: IBGE (2023).

The result of animal species assessments rose from 12,009 in 2014 to 13,939 in 2022 in the Red Book (Tabela 13). Of the total assessed in 2022, 1,253 animal species were in the Critically Endangered (CR), Endangered (EM) or Vulnerable (VU) categories, i.e. around 9.0% of the species assessed were at risk of extinction.

TABELA 13 - Species by category in the 2014 and 2022 assessments

Risk Categories	Species by taxonomic group, habit or life form per year							
	Fauna - total				Flora - total			
	2014	%	2022	%	2014	%	2022	%
Extinct	10	0,1%	9	0,1%		0,0%		0,0%
Extinct in the wild	1	0,0%	1	0,0%		0,0%		0,0%
Critically endangered	318	2,6%	358	2,6%	446	10,4%	684	9,1%
In danger	405	3,4%	427	3,1%	1112	25,8%	1843	24,5%
Vulnerable	448	3,7%	468	3,4%	481	11,2%	680	9,0%
Almost threatened	314	2,6%	289	2,1%	257	6,0%	451	6,0%
Less worrying	8845	73,7%	11278	80,9%	1535	35,7%	2612	34,7%
Insufficient data	1668	13,9%	1109	8,0%	473	11,0%	1247	16,6%
Total	**12009**	**100,0%**	**13939**	**100,0%**	**4304**	**100,0%**	**7517**	**100,0%**

Source: IBGE (2023).

The majority of threatened animal and plant species in the 2022 assessment are found in the Atlantic Forest, around 24% (Figura 84 e Tabela 14), followed by the Cerrado (16%) and the Caatinga (15%), then the Pampa (10%, the Coastal Marine System (7%) and the Pantanal (4%).

Figura 84 - Percentage of threatened species by biome in 2022, in relation to the total number of threatened species of animals and plants.

TABELA 14 - Species assessed and threatened by biome per year

Biome	2014			2022		
	Species evaluated	Species threatened	%	Species evaluated	Species threatened	%
Atlantic Forest	9042	2016	22	11811	2845	24
Amazon	6515	311	5	8346	503	6
Cerrado	6290	1037	16	7385	1199	16
Caatinga	2714	395	15	3220	481	15
Coastal Marine System	2091	166	8	2286	170	7
Pampa	2011	234	12	2185	229	10
Pantanal	1718	65	4	1826	74	4
Total	**30381**	**4224**	**82**	**37059**	**5501**	**82**

Source: IBGE (2023).

9 FAUNA INVENTORY

A fauna inventory is the qualitative or quantitative description of a population or community of animals in a given location over a defined period of time.

The aim of inventories is to detect populations of organisms in a given sampling space and time period with the aim of conserving and managing the population or community being assessed.

Human beings are at the top of food chains and depend on the health and balance of all species for their well-being and survival. Natural spaces are increasingly reduced and affected by anthropogenic actions, making it necessary to assess the state of natural ecosystems for the conservation and management of natural communities. The biodiversity of a natural community is highly influenced by fauna, as animals carry out various activities that are essential to the life of all biota, such as turning over and fertilizing the soil, dispersing seeds and pollinating plants.

Fauna inventories help to assess the health and stability of ecosystems and to prevent and remedy problems caused by anthropogenic actions and natural disasters. The main reasons for carrying out fauna inventories are:

> Conflicts with crops or livestock;
> Occurrence of disease transmitters;
> Population declines of species useful to humans;
> Conflicts over excessive population growth of certain species;
> Invasions of exotic species;
> Economic, social, cultural and educational value;
> Implementation and management of conservation units;
> Natural area management for economic production;
> Planning by an official agency or non-governmental organization;
> Need for environmental recovery;
> Legal requirement for environmental licensing;
> Legal requirement for the operation of rural enterprises.

For more accurate fauna studies, due to the constant movement of animals, there is a need for adequate monitoring of each species, in order to determine their presence in different seasons and in different years.

The aim of fauna monitoring is to catalog the species that exist in a given region, evaluating the etiology in relation to survival, reproduction, migration of specimens to other habitats, etc. During the implementation and operation phases of economic ventures, fauna monitoring should be carried out every three months, lasting six to eight days for each field survey (CRMV-SP, 2023).

Sampling is the collection of data from a representative portion of a population or community.

Fauna sampling is a technical-scientific endeavor whose procedures must be guided by sampling theory and must be statistically based, regardless of the class of animal being inventoried.

Qualitative inventories seek to list the species occurring in an area during a given period in order to determine species richness.

Quantitative inventories generally seek to determine the biodiversity of an area over a given period.

The classes of animals and ecosystems will determine the methods to be used, i.e. the type of sample itself. The methods used to detect and count fish are different from those used for mammals or birds. For each animal species, the appropriate method must be chosen for its detection, i.e. different types of samples are used for each type of animal and ecosystem.

Attention: the distribution of sampling units must always be statistically based and without human interference in the choice of where the sampling units will be located, in order for the sampling to be scientifically valid.

9.1 Sampling theory

Sampling is the assessment of a population or community from a fraction of the total area occupied by the population or community under study.

Main characteristics of sampling:

- ➢ Sample variables and attributes;
- ➢ Sample statistics;
- ➢ Sampling intensity;
- ➢ Type of units and sampling methods;
- ➢ Sampling systems;
- ➢ And sampling processes.

9.1.1 Variables

These are the characteristics obtained about the animals in each sampling unit, such as:

- ➢ Place of observation;
- ➢ Type of habitat;
- ➢ Date and time of observation;
- ➢ Species observed;
- ➢ Age class (size or stage of development);
- ➢ Data for individual identification (color, spots, rings, etc.);
- ➢ Activity;
- ➢ Number of individuals per species;
- ➢ Sound;
- ➢ Climate data (temperature, rainfall, cloudiness).

9.1.2 Sample Statistics

They can only be estimated when there are repeated observations in the sampling units. The statistics are calculated for the population studied on the variables observed in the sampling units, mainly:

- Average ($\bar{Y}$);
- Median (Me);
- Fashion (Mo);
- Minimum Limit (LI) and Maximum Limit (LM);

- Variance (S^2);
- Standard deviation (S);
- Coefficient of Variation (CV).

The statistics will result in characteristics such as:

- ➢ Number of species (richness);
- ➢ Population density (individuals per species per unit area);
- ➢ Distribution by age class (puerile, young, mature, senile);
- ➢ etc.

9.1.3 Intensity and sampling error

Sampling intensity is the ratio of sample size to population size. The sampling intensity must be such that it includes all or almost all of the species found on the site.

The sampling units must not overlap and must be distributed throughout the area to be inventoried. It is necessary to randomly or systematically distribute the locations for installing the sampling units, so that they all have the same probability of being chosen.

Ideally, repeated observations in the sampling units should include all the times and seasons of occurrence of the different species.

The sampling error will be greater the less intense the sampling and should be established before sampling to determine the number of sampling units required. The natural variation that exists between sites and habitats is a characteristic that must be taken into account in fauna inventories.

The statistics are calculated in the same way as those used in forest inventories and the calculation methodology will depend on the type and distribution system of the sampling units and the sampling process used.

9.1.4 Type of sampling units and methods

The type of sampling units and methods to be used depend on the type of animal to be observed. Sampling units can be fixed area or variable area. Fauna sampling methods can be by:

- ➢ Active search;
- ➢ Direct evidence;
- ➢ Indirect evidence;
- ➢ Vehicle search;
- ➢ Casual encounters.

9.1.4.1 Active search

Active search for individuals, tracks and traces are searched for by going around the study area, looking for animals, footprints, feces, carcasses, fur, burrows and other signs of occurrence. In the active search, the search points (sampling units) cannot be chosen at random, they need to be distributed randomly (draw) or systematically in the area to be inventoried.

9.1.4.2 Direct evidence

In the observation of direct evidence (Figura 85) it is necessary to distribute the sites (sampling units) randomly or systematically in the area to be inventoried.

Figura 85 - Direct detection methods - traps.

Figura 86 - Direct detection methods - traps.

Figura 87 - Direct detection methods - observation tower.

9.1.4.3 Indirect evidence

When observing indirect evidence, it is necessary to distribute the sites (sampling units) randomly or systematically in the area to be inventoried.

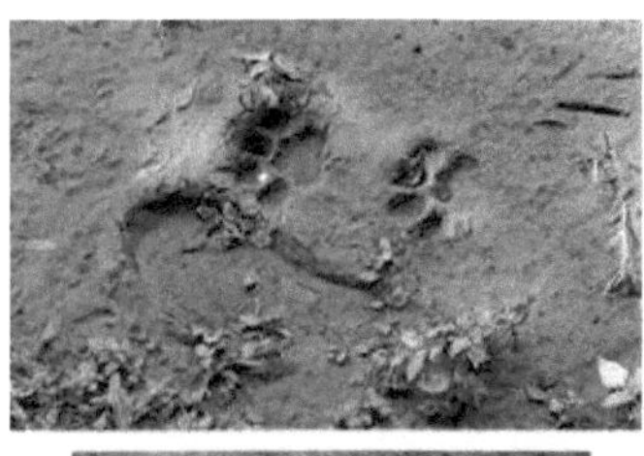

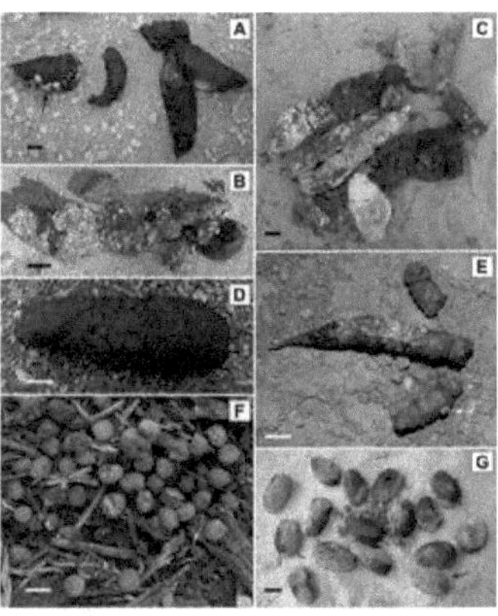

Figura 88 - Indirect detection methods - footprints and droppings.

9.1.4.4 Vehicle search

> It is a type of search for direct visual and auditory evidence on board vehicles;
> It should be done at low speed (<40km/h);
> Roads, tracks and access routes are traveled within and between different phytophysiognomies;
> When possible, photographs, sound recordings and videos can be taken.

9.1.4.5 Casual encounters

In this type of encounter, the animals observed must be identified and counted. It is not used to compile statistics, only to record the occurrence of species and the occurrence of flocks, packs, herds, etc.

9.1.5 Identification and identity of the animals

Animal identification should preferably be carried out by specialists. Scientific names should be obtained from official lists and specialized literature. Whenever possible, the animals should be photographed or drawn for later correct identification and important aspects should be noted, such as colors and patterns of spots, type and size of structures such as fur, crests, beaks, feet, horns, among others that can help with identification. Footprints, feathers and fur can also be used for identification (Figura 89).

Figura 89 - Guides for identifying animals by their fur and footprints.

The individual identity of the animals must be verified so that they are not counted more than once each time and recapture determined. In addition to having their individual identity verified by color patterns, spots, differences in structures such as fins, animals can also be: ringed or chipped; tattooed; have their toes cut off; have seals, earrings or earrings attached; ear pricks and ear holes using the Australian system; marks that allow individual identity to be verified and avoid double counting in population density studies, as well as verifying recaptures in different places and at different times for migration and growth studies.

9.2 Detection methods by animal class

Some detection methods are listed below for the following categories of animals:
- Ichthyofauna;
- Herpetofauna;
- Birdlife;
- Mastofauna;
- Carcinofauna;
- Malacofauna;
- Entomofauna.

Records of animals observed or captured must include:
- Complete taxonomic identification (genus and species when possible);
- Photographs and videos catalogued;
- Recording vocalizations (sound recordings);
- Place of observation;
- Type of habitat;
- Date and time of observation;
- Species observed;
- Age class (vital stage: puerile, young, mature, senile);

➢ Data for individual identification (color, spots, rings, etc.);

➢ Activity;

➢ Number of individuals per species;

➢ Climate data (temperature, rainfall, cloudiness).

9.2.1 Detection methods for ICTIOFAUNA

The ichthyofauna is the group of fish (Figura 90) that live in a given place.

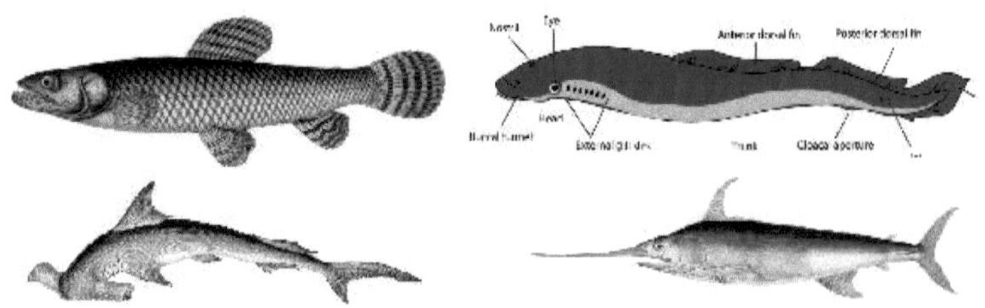

Figura 90 - Fish from different ecosystems.

9.2.1.1 Active search

It is carried out with the help of:

➢ traps;

➢ shovels;

➢ tarrafas;

➢ fishing nets;

➢ Diving with a camera;

➢ Sonar for detecting shoals and large fish;

➢ Baits to attract fish.

9.2.1.2 Materials for ichthyofauna surveys

➢ Traps for fish and crustaceans;

➢ Diving flag;

➢ Boat;

➢ Buoys;

➢ Underwater photography and video camera;

- ➢ Field file;
- ➢ Wellies;
- ➢ GPS;
- ➢ Fish lures;
- ➢ Diving kit;
- ➢ Scale;
- ➢ First aid kit;
- ➢ Fishing equipment in general;
- ➢ Sieve for shrimp, tadpoles and fry;
- ➢ Underwater clipboard;
- ➢ Sunscreen;
- ➢ Puçá;
- ➢ Waiting network;
- ➢ Repellent;
- ➢ Water shoes;
- ➢ Sonar;
- ➢ Tarrafa;
- ➢ Trena.

9.2.2 Detection methods for HERPETOFAUNA

Herpetofauna is the group of amphibians and reptiles (Figura 91) that live in a particular place.

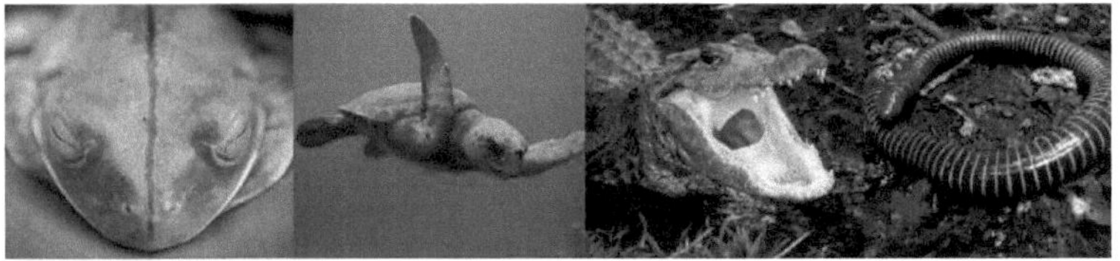

Figura 91 - Amphibians and reptiles from different ecosystems.

9.2.2.1 Active search

- ➢ Day and night with visual and auditory recording;
- ➢ Use a flashlight;

> Search on vegetation, in leaf litter, tree tops and hollows, on the ground, leaf litter, under rocks, trunks, potential shelters, bromeliads and breeding sites.
> search for traces and tracks.

9.2.2.2 Materials for surveying herpetofauna

> Video and photo camera;
> Recorder;
> Binoculars;
> Flashlight;
> Field file;
> GPS;
> Puçá;
> Tweezers;
> Hook;
> Boots;
> Leg;
> Sunscreen;
> First aid kit.

9.2.3 Detection methods for AVIFAUNA

Avifauna is the group of birds (Figura 92) that inhabit a particular place.

Birdwatching times should preferably be in the early hours of the morning, late afternoon and at night.

Figura 92 - Birds from different ecosystems.

9.2.3.1 Birdlife can be inventoried using the following methods:

- ➢ Species Count by Time (CET);
- ➢ Mackinnon lists.

9.2.3.2 The main ways of detecting birdlife are:

- ➢ Search for traces such as footprints, feathers, nests, regurgitations.
- ➢ Nocturnal birds should be searched for with flashlights;
- ➢ The use of recorded sound recordings and a digital recorder to record responses are useful for detection and identification.

9.2.3.3 Materials for bird surveys:

- ➢ Video and photo camera;
- ➢ Binoculars;
- ➢ Flashlight;
- ➢ Recorder;
- ➢ Speaker;
- ➢ GPS;
- ➢ Field sheets;
- ➢ Field guides;
- ➢ Leggings;
- ➢ Boots;
- ➢ Repellent;
- ➢ Sunscreen;
- ➢ First aid kit.

9.2.4 Detection methods for MASTOFAUNA

Mastofauna is the group of mammals (Figura 93) that live in a particular place.

Figura 93 - Mammals from different ecosystems.

It is necessary to separate mammals in terms of habitat and feeding habits:

- ➢ Aquatic mammals:
- ➢ Marine;
- ➢ Freshwater;
- ➢ Land mammals:
  - o Herbivores;
  - o Carnivores;
  - o Omnivores;
- ➢ Arboreal mammals;
- ➢ Chiroptera.

9.2.4.1 Detection methods for mastofauna

Camera traps (active during the day and/or night) can be used to detect mastofauna, with or without bait.

In the active search for mastofauna, we look for trails, tracks, footprints, burrows, feces, fur, gnawed or skinned bark and carcasses.

Mastofauna are also used:

- ➢ Direct visualization at sampling points.
- ➢ Vocalizations emitted for response and recording.
- ➢ Traps in general.
- ➢ Mist nets for chiropters.

9.2.4.2 Materials for mastofauna surveys:

- ➢ Video camera, sound recorder and photographic camera;
- ➢ Camera traps;

> Trap door;

> Bait;

> Binoculars;

> GPS;

> Field sheets;

> Leggings;

> Boots;

> Repellent;

> Sunscreen;

> First aid kit.

9.2.5 Detection methods for CARCINOFAUNA

Carcinofauna is the group of barnacles, amphipods, isopods, shrimps, crayfish, lobsters and crabs (Figura 94) that live in a given location.

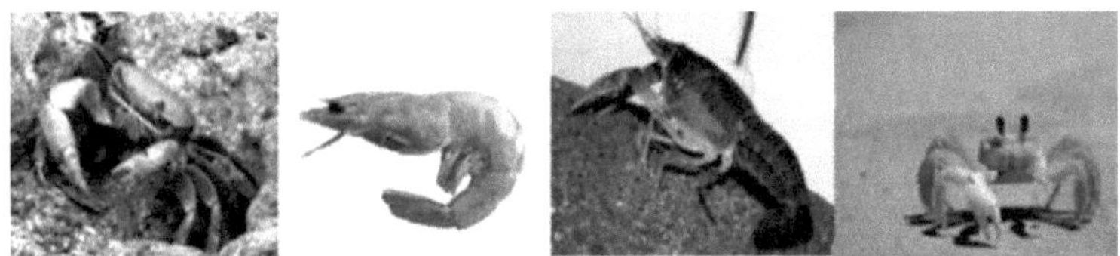

Figura 94 - Species of carcinofauna from different ecosystems.

The most common methods for detecting shrimp and shrimp-like creatures are catching them with baskets, sieves and traps, usually with bait.

9.2.6 Detection methods for MALACOFAUNA

Malacofauna is the group of molluscs (Figura 95) that inhabit a given location.

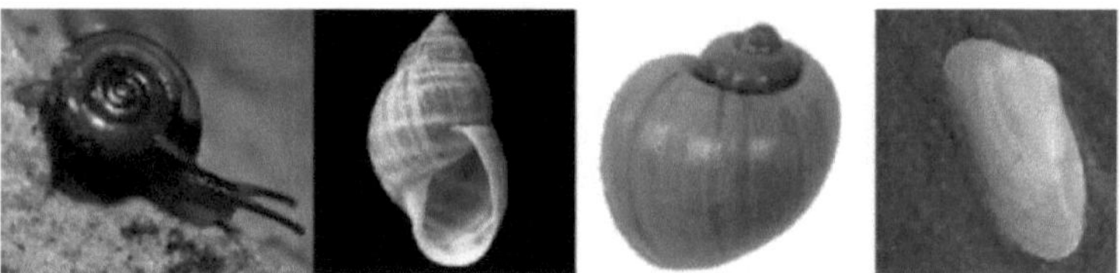

Figura 95 - Molluscs from different ecosystems.

Malacofauna is made up of the classes *Caudofoveata, Aplacophora, Polyplacophora, Monoplacophora, Bivalvia, Scaphopoda, Gastropoda* and Cephalopoda and the following three are the most common to be included in inventories:

> Gastropods - single shell - can burrow or not;
> Bivalves - 2 articulated shells - buried (massunim) or anchored (sururu);
> Cephalopods - do not have an external shell.

9.2.7 Detection methods for ENTOMOFAUNA

Entomofauna is the group of insects (Figura 96) that live in a particular place.

Figura 96 - Insects from different ecosystems.

Detection methods differ for each class, generally by active search without tools or utensils, or using sieves, shovels and baited traps.

They can be used:

> Traps with bait, pheromones or light;
> Nets for catching insects;
> Camera;

> Active search for adults, larvae and pupae.

9.2.8 Example of a field sheet

In field surveys, in each sampling unit, it is recommended to record data on the site and ecosystem, common name and individual identity, scientific name, vital stage and activities practiced by the animals, using a spreadsheet similar to the one shown in Table 1 below:

Table 1 - Example of a spreadsheet for data collection in a fauna inventory

Location:		Coordinates:		Sample Unit:	Ecosystem:			Time:				Date and time:		
Common name / identity		Scientific name		Life stage	Activities									
					AL	AC	CH	CN	AP	PD	FU	MI	PR	OR
...														
Life stage: Child / Youth / Adult	ACTIVITIES: | AL - Feeding | AC - Mating | CH - Brooding | CN - Nest/shelter building | AP - Offspring feeding | PD - Predation | FU - Escape | MI - Migration | PR - Offspring protection |													
OR - Other Activities:														

9.2.9 Sample design

It is defined by the sampling system adopted. It is the way in which the sampling units are distributed and this implies a specific statistical approach for each system. It must be linked to the objective and take into account the particularities of the organisms. For the sample to be representative of the population, it must have a valid statistical design and take into account the seasonality of the populations.

9.2.10 Sampling effort

Delimiting the sampling effort in advance is essential for achieving the objectives of the inventory, such as:

- ➢ Establish the number of sampling units (standardization);
- ➢ Observation period;
- ➢ Data collection exposure time.

Some of the methods for collecting data on animals are as follows:

- ➢ Direct observations;
- ➢ Traps;
- ➢ Indirect evidence;
- ➢ Auditory search.

9.2.11 Data processing

Planning the treatment of the data to be collected is an important part of the inventory, and prior studies should also be carried out:

- ➢ Draw up a preliminary list of animals occurring in the inventory region based on databases and bibliography;
- ➢ Choose indices and estimators without exaggeration;
- ➢ Determine the form of interpretation.

9.2.12 Biological variables

When choosing the biological variables to record in the survey, consideration should be given:

- ➢ The evolutionary history of species;
- ➢ The physiology of species;
- ➢ The behavior of species (ethology).

9.2.13 Sampling systems

They characterize the way in which the sampling units are distributed and the statistical approach; the main systems for distributing the sampling units are the following five:

> Unrestricted random - The sampling units are distributed by lot over the inventory area;

> Random with restrictions - The population is divided into strata according to the occurrence of different phytophysiognomies or ecosystems;

> Systematic without restrictions - The sampling units are distributed equidistantly in the area to be inventoried;

> Systematic with restrictions - The area is divided into strata (green, yellow, orange, red) according to the occurrence of different phytophysiognomies and the sampling units are distributed systematically;

> Mixed (part random and part systematic) - This is restricted sampling in which the distribution of sampling units is random at one level and systematic at another.

9.2.14 Sampling procedures

Sample sufficiency in any sampling process must first take into account the stabilization of the curve of the number of species as a function of the increase in the sampling area, or the increase in the number of sampling units, known as the species/area curve. When the curve stabilizes, without increasing the number of species found in the survey, it is considered that the sample has reached a sufficient size. The curve can also be constructed as a function of the number of species found in relation to the sampling time (Figura 97) and also the number of species found in relation to the number of sampling units.

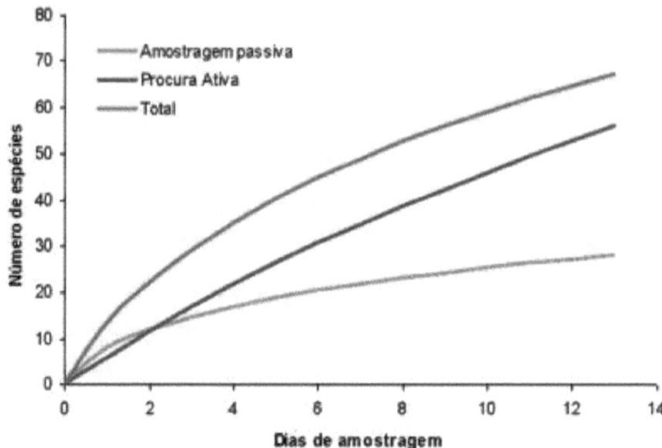

Figura 97 - Randomized species accumulation curve (1,000 randomizations) based on the herpetofauna sampled over 13 days in the Cerrado of Central Brazil. Source: Silveira et al (2010).

Sampling processes are categorized by the type of sampling unit, sampling system and form of distribution.

The most common processes are:

> AAS - simple random sampling;
> SEA - stratified random sampling;
> ASL - systematic line sampling;
> ASF - systematic strip or transect sampling;
> ASR - systematic sampling in networks;
> AC - cluster sampling;
> AME - multi-stage sampling.

9.2.14.1 AAS - Simple random sampling

Sampling units are distributed by lot over the inventoried area as shown in Figura 98.

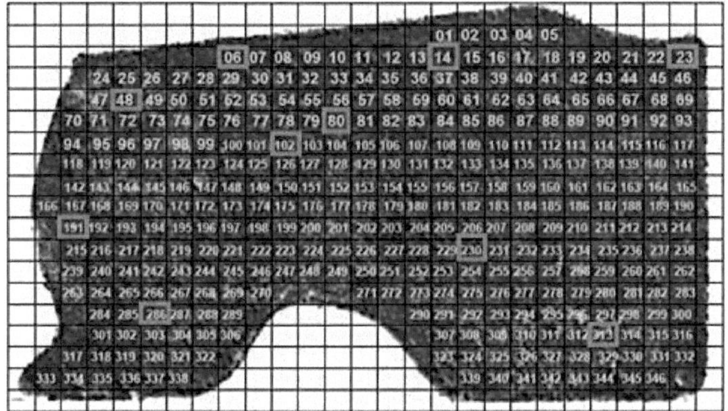

Figura 98 - Simple random sampling, with units 06, 14, 23, 48, 80, 102, 191, 230, 286 and 313 drawn to make up the sample.

In AAS, the statistical approach is characteristically unrestricted, as follows.

Sampling intensity

Sampling Intensity (SI) is represented by the percentage of the study area that was sampled. The AI is calculated by:

$$AI = 100 . n . a / A$$

Where: AI = Sampling Intensity in percentage; n = number of sampling units; a = area of the sampling unit in square meters; A = area of the population in square meters.

As an example, consider a forest plantation with an area of 100 hectares, or 1,000,000 m², subjected to a sample of 50 sampling units of 400 m² each, making up a sample of 50 x 400 m² = 20000 m², or 2 hectares; the sample therefore represents a Sampling Intensity (SI) of 2% of the population area, which makes it a finite population.

The sampling intensity can be calculated alternatively by the percentage of sampling units used to represent the population (n) and the potential number of sampling units (N). The potential number of sampling units (N) is calculated by dividing the total area of the population by the area of a single sampling unit. The sampling intensity

Populations whose sampling intensity is less than 2% are said to be infinite. When the sampling intensity is 2% or more, the population is said to be finite. Finite populations need to have their statistics corrected to reduce bias, using a multiplicative correction factor (CF) calculated by the expression:

$$FC = \left(\frac{N - n}{N}\right)$$

Where: FC = correction factor; n = number of sampling units; N = potential number of sampling units in the population.

The formulas for finite populations can be considered as generalized and can be used for both cases, since the correction factor becomes insignificant in infinite populations. The purpose of the correction factor is to reduce bias and it exists even in infinite populations

Average

Sample averages are calculated using the expression:

$$\bar{y} = \frac{\sum_{i=1}^{n} \bar{y}_i}{n}$$

Where: $\bar{y}$ = sample mean; $\bar{y}_i$ = mean of sample unit i; n = number of sample units; i = order number of sample unit.

Variance

The sample variance is calculated by:

$$s_x^2 = \frac{\sum_{i=1}^{n}(\bar{y}_i - \bar{y})^2}{n - 1}$$

Where: s_y^2 = sample variance; $\bar{y}$ = sample mean; $\bar{y}_i$ = mean of sample unit i; n = number of sample units; i = order number of sample unit.

Sample sufficiency (n)$_s$

Sample sufficiency must take into account the stabilization of the species/area curve.

Sample sufficiency is defined as the number of sample units sufficient to represent the population with the maximum error limit allowed, previously determined in relation to the population mean. The calculation is carried out for an infinite population without the correction factor and for a finite population with the correction factor, as follows:

> ➢ For infinite population

$$n = \frac{t^2_{(p,GL)} \cdot s_y^2}{E^2}$$

> ➢ For finite population

$$n = \frac{N \cdot t^2_{(p,GL)} \cdot s_y^2}{N \cdot E^2 + t^2_{(p,GL)} \cdot s_y^2}$$

Where: n = number of sample units sufficient to represent the population; s_y^2 = sample variance; $\bar{y}$ = sample mean; LE = allowed sampling error limit (LE% x Mean); N = total potential number of sampling units; t(p,GL) = Student's t-value for n-1 degrees of freedom (GL) and probability of error p; n = number of sampling units.

Coefficient of variation

The coefficient of variation as a percentage of the mean is calculated using the expression:

$$cv = \frac{100 \cdot s_y}{\bar{y}}$$

Where: CV = coefficient of variation ; s_y = standard deviation; $\bar{y}$ = sample mean.

Variance of the mean

The Variance of the Mean is determined by:

- ➢ For infinite population

$$S_{\bar{y}}^2 = \pm \frac{S_y^2}{n-1}$$

- ➢ For finite population

$$s_{\bar{y}}^2 = \left(\frac{s_y^2}{n-1}\right) \cdot \left(\frac{N-n}{N}\right)$$

Where: $S_{\bar{y}}^2$ = variance of the mean; S_y^2 = variance; n = number of sample units; N = potential number of sample units.

Standard error of the mean

The Standard Error of the Mean is determined by:

- ➢ For infinite population

$$S_{\bar{y}} = \pm \frac{s_y}{\sqrt{n}}$$

- ➢ For finite population

$$S_{\bar{y}} = \pm \frac{s_y}{\sqrt{n}} \sqrt{\left(\frac{N-n}{N}\right)}$$

Where: $S_{\bar{y}}$ = standard error of the mean; s_y = standard deviation; n = number of sampling units; N = potential number of sampling units.

Sample error

The Sampling Error is calculated for absolute and relative values, as follows:

- ➢ Absolute sampling error (Ea)

$$E_a = \pm \, t_{(p,GL)} \cdot S_{\bar{y}}$$

- ➢ Relative sampling error (Er)

$$E_r = \pm \ \frac{100.t_{(p,GL)}.S_{\bar{y}}}{\bar{y}}$$

Where: Ea, Er = absolute and relative sampling error, respectively; $S_{\bar{y}}$ = standard error of the mean; $t_{(p,GL)}$ = Student's t-value for n-1 degrees of freedom (GL) and probability of error (p); GL = n - 1 (degrees of freedom); p = probability of error (p=1-P).

Confidence interval

$$IC\ [(\bar{y} - \ t_{(p,GL)}.S_y) \leq \mu \leq (\bar{y} + \ t_{(p,GL)}.S_{\bar{y}})] = P$$

Where: CI = confidence interval; $\bar{y}$ = sample mean; μ = population mean; $S_{\bar{y}}$ = standard error of the mean; $t_{(p,GL)}$ = Student's t-value for n-1 degrees of freedom (GL) and probability of error (p=1-P);
P = probability of confidence for the mean.

Minimum confidence estimate

$$EMC = \bar{y} - \ t_{(p,GL)}.S_{\bar{y}}$$

Where: EMC = minimum confidence estimate; $\bar{y}$ = sample mean; $S_{\bar{y}}$ = standard error of the mean; $t_{(p,GL)}$ = Student's t-value for n-1 degrees of freedom (GL) and probability of error (p).

9.2.14.2 SEA - Stratified random sampling

The population is divided into strata according to the occurrence of different phytophysiognomies, as in the example in Figura 99with 3 strata, represented by different colors. The smallest stratum represents around 15% of the total area, the largest around 50% and the intermediate stratum around 35% of the total area to be inventoried. 20 sampling units (AU) were distributed in the strata proportionally, i.e. 3 AU, 7 AU and 10 AU, respectively.

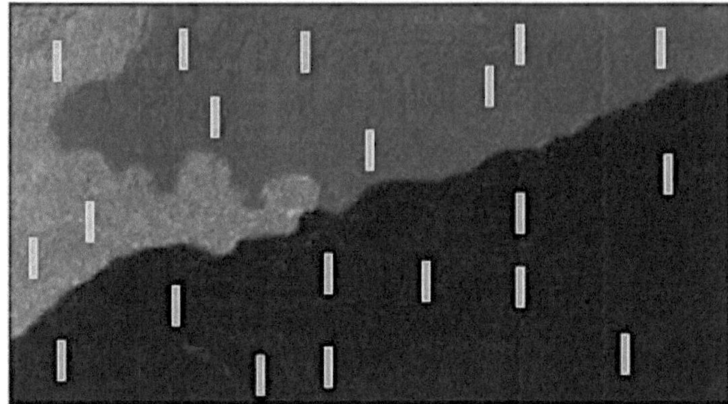

Figura 99 - Sampling in three strata, with random distribution of 20 sampling units in proportion to the area of each stratum.

In SEA, the statistical approach is carried out using the system with restrictions, whereby the sampling area is usually divided only into strata represented by the existing ecosystems, i.e. there is only one restriction, which is to divide the area only once into strata represented by the different ecosystems.

When the area to be inventoried is very large, or has different types of vegetation, there can be a great deal of variation from one site to another, which could mean a very large number of sampling units for the required precision. In order to reduce the variation between sampling units, sampling with division into strata is used.

Very large areas can be divided into large 1st level sampling units and these subdivided again into smaller 2nd level sampling units and these can again be subdivided into smaller 3rd level units and so on.

In restricted sampling, there are cases in which the 1st level units have different sizes, so it is necessary to calculate the means and variances considering the proportionality factor, or weight (w_h) of each one, which is calculated by:

$$w_h = A_h / A$$

Where: w_h = weight or proportion of the area of level h over the total; A_h = area of the 1st level unit of order h considered; h = order number of the 1st level unit; A = Total area of the population, obtained by adding up the areas of all the 1st level units.

If the areas of the 1st level units are the same size, the weight w_h will be the same for all 1st level units.

9.2.15 Sampling with a restriction

Population division levels for sampling with a restriction

Consider a forest area to be inventoried like the one shown in Figura 100. Dividing the area of Figura 100 into 9 1st level units, whose order numbers are represented by the letter h.

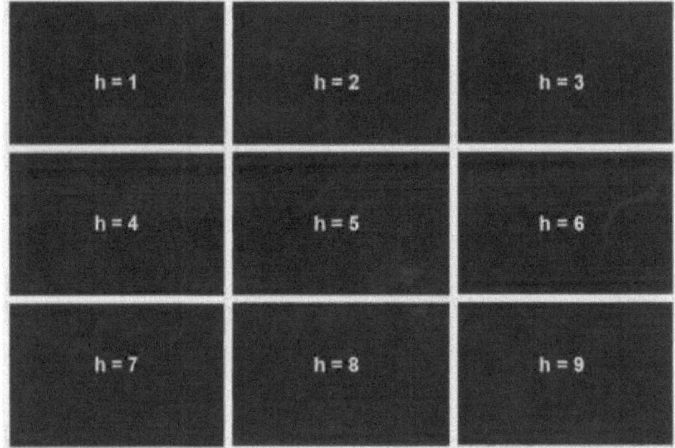

Figura 100 - Forest area to be inventoried with representation of 9 1st level units, where h is the order number of the unit in the sampling.

The total number of 1st level units in the population is represented by the letter H, H=9 in this case, and their order number is h. The number of 2nd level sample units within each 1st level unit is represented by n_h and the order number of each is represented by the letter i.

Sample mean with a restriction

$$\bar{\bar{y}} = \sum_{h=1}^{H} w_h \, \bar{y}_h$$

Where: $\bar{\bar{y}}$ = average of 1st level units; $\bar{y}_j$ = average of the 1st level unit of order h; w_h = proportion of the area of the 1st level unit of order h over the total area; H = sample number of 1st level units; h = order number of the 1st level unit.

1st level unit averages

$$\bar{y}_h = \frac{\sum_{i=1}^{n_h} \bar{y}_{hi}}{n_h}$$

Where: $\bar{y}_h$ = average of the 1st level unit of order h; $\bar{y}_{hi}$ = average of the 2nd level unit of order i; n_h = number of 2nd level units in the 1st level unit of order h; h = order number of the 1st level unit; i = order number of the 2nd level unit.

Sample variance

$$s^2 = \sum_{h=1}^{H} w_h \, s_h^2$$

Where: s^2 = variance; $s_h{}^2$ = variance of the 1st level unit of order h; w_h = proportion of the area of the 1st level unit of order h over the total area; H = sample number of 1st level units; h = order number of the 1st level unit.

Sample sufficiency (n)

Infinite population

$$n = \frac{t_{(p,GL)}^2 \, \sum_{h=1}^{H} w_h \cdot s_h^2}{E^2}$$

Finite population

$$n = \frac{N . t_{(p,GL)}^2 \, \sum_{h=1}^{H} s_h^2}{N.E^2 + t_{(p,GL)}^2 \, \sum_{h=1}^{H} w_h \cdot s_h^2}$$

Where: s_h^2 = variance of the 1st level unit of order h; LE = admitted sampling error limit; N = total potential number of sampling units in the population; H = number of 1st level units; w_h = proportion of the area of the 1st level unit of order h over the total area; $t_{(p, GL)}$ = Student's t-value for n-1 degrees of freedom (GL) and probability of error (p).

Allocation of AUs

Once the sample sufficiency (n) has been determined, the AUs can be allocated to each 1st level unit solely by the area proportion (w_h), or by the proportion multiplied by the standard deviation (w_h s)$._h$

Allocation in proportion to the area of each 1st level UA:

$$n_h = n \ w_h$$

Where: n_h = number of sampling units to be allocated to the 1st level unit of order h; n = number of total sampling units needed in the final inventory; w_h = weight or area proportion of the 1st level unit of order h.

Allocation of sampling units carried out in each 1st level AU by the proportion of area (wh) and standard deviation (sj) of each 1st level unit:

$$n_h = \frac{n \ w_h \ s_h}{s}$$

Where: n_h = number of sampling units to be allocated to the 1st level unit of order h; n = number of total sampling units needed in the definitive inventory; w_h = area proportion of the 1st level AU of order h; s_h = standard deviation of the 1st level AU of order h; s = sampling standard deviation of the stand.

Analysis of sampling variance *with one restriction*

TABELA 15 - Analysis of sampling variance with one restriction

Variation Factor	Degrees of Freedom	Sum of Squares	Mean Square	F
Between 1st level UAs	H - 1	$\sum_{h=1}^{H} n_h (\bar{y}_h - \bar{\bar{y}})^2$	SQe / GLe	QMe / QMd
Within the 1st level UAs	n - H	$\sum_{h=1}^{H} \sum_{i=1}^{n_h} (\bar{y}_{hi} - \bar{y}_h)^2$	SQd / GLd	-
Total	n - 1	$\sum_{h=1}^{H} \sum_{i=1}^{n_h} (\bar{y}_{hi} - \bar{\bar{y}})^2$		-

Interpretation of analysis of variance:

> ➤ If the F value is significant: there is a difference between the 1st level units and the division is efficient, reducing the sampling error;

> ➤ If the F value is not significant: there is no difference between the 1st level units and the division is unnecessary.

Coefficient of *variation*

$$CV = \frac{100\, s}{\bar{\bar{y}}}$$

Where: CV = population coefficient of variation; $\bar{\bar{y}}$ = sample mean; s = sample standard deviation. Statistics with restrictions (2 levels)

Variance of *the mean*

Infinite population

$$s_{\bar{y}}^2 = \sum_{h=1}^{H} \frac{w_h^2 \cdot s_h^2}{n_h}$$

Finite population

$$s_{\bar{\bar{y}}}^2 = \sum_{h=1}^{H} \frac{w_h^2 . s_h^2}{n_h} - \sum_{h=1}^{m} \frac{w_h . s_h^2}{N}$$

Where: $s_{\bar{\bar{y}}}^2$ = variance of the population mean; s_h ^{2} = variance of the 1st level unit of order h; w_h = proportion of the area of the 1st level unit of order h over the total area; H = number of 1st level units; n_h = number of 2nd level units in the 1st level unit of order h; N = potential number of sample units from the whole population; h = order number of the 1st level unit.

Sample error

$$E_a = \pm \ t_{(p,GL)} . S_{\bar{\bar{y}}} \qquad E_r = \pm \ \frac{100 . t_{(p,GL)} . S_{\bar{\bar{y}}}}{\bar{\bar{y}}}$$

Where: Ea, Er = absolute and relative sampling error, respectively; $\bar{\bar{y}}$ = population mean; $S_{\bar{\bar{y}}}$ = standard error of the population mean; $t_{(p, GL)}$ = Student's t-value for n-1 degrees of freedom (GL) and probability of error (p).

Confidence interval

$$IC \ [(\bar{\bar{y}} - \ t_{(p,GL)} . s_{\bar{\bar{y}}}) \leq \mu \leq (\bar{\bar{y}} + \ t_{(p,GL)} . s_{\bar{\bar{y}}})] = P$$

Where: CI = confidence interval; $\bar{\bar{y}}$ = sample mean; μ = population mean; $s_{\bar{\bar{y}}}$ = standard error of the population mean; P = probability of confidence for the mean; GL = degrees of freedom = number of strata-1; p = admitted probability of error = 1-P; $t_{(p, GL)}$ = Student's t-value for n-1 degrees of freedom (GL) and admitted probability of error (p); P = probability of confidence for the mean.

Minimum confidence estimate

$$EMC = \bar{\bar{y}} - \ t_{(p,GL)} \ s_{\bar{\bar{y}}}$$

Where: CME = minimum confidence estimate; $\bar{\bar{y}}$ = population mean; $s_{\bar{\bar{y}}}$ = standard error of the population mean; p = probability of error; GL = degrees of freedom (number of strata - 1); $t_{(p,GL)}$ = Student's t-value for n-1 degrees of freedom (GL) and probability of error (p).

9.2.15.1 ASL - systematic line sampling

The SLA can use the same statistical approach as the SEA, where each line will represent the sampling units of a stratum. The strata will be divided by a straight line down the middle of two sampling lines, and the area of each stratum formed in this way will be measured to determine the weights for calculating the stratified mean and variance.

The lines and sampling units are distributed equidistantly over the area to be inventoried (Figura 101).

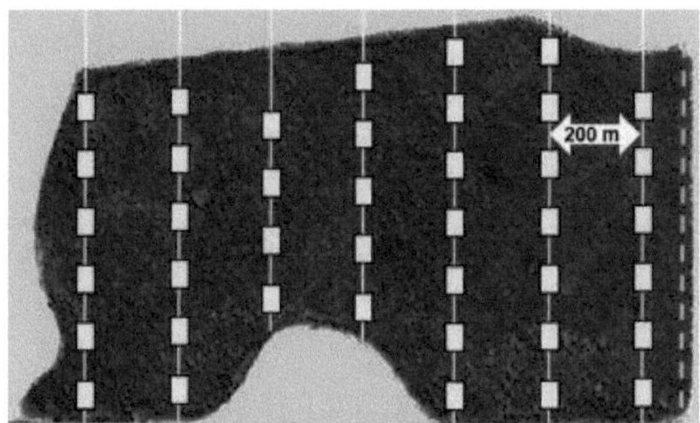

Figura 101 - Systematic distribution of sampling units in lines 200 m apart.

9.2.15.2 ASF - systematic strip or transect sampling

A belt or strip transect is a long rectangular plot of fixed width (Figura 102), crossing the entire vegetation type under study. They are very useful for studies of natural vegetation in small forest areas up to a maximum of 500 hectares, with systematically distributed sampling units.

In ASF, the statistical approach is similar to ASL and the division of strata will be determined by dividing the area by lines in the middle of two sampling bands. Likewise, the area of each stratum must be calculated to determine the proportion of each stratum (weight) so that the stratified mean and variance can be calculated.

Figura 102 - Systematic sampling in equidistant transects.

9.2.15.3 ASR - systematic sampling in networks

The sampling units are distributed in a network of equidistant points (Figura 103).

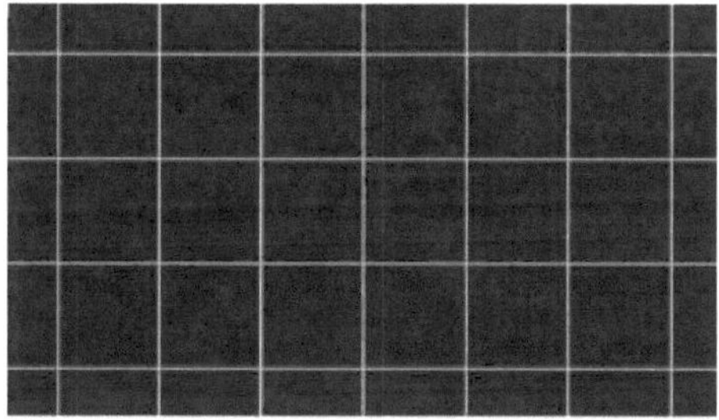

Figura 103 - Distribution of the sampling units in a network of 28 equidistant points on the intersections of the lines.

In the case of ASR, the statistical approach can be the same as for AAS, since the weight of each sampling unit is the same for all of them.

9.2.15.4 AC - cluster sampling

The sampling units are distributed in groups of plots of standardized size and shape (Figura 104).

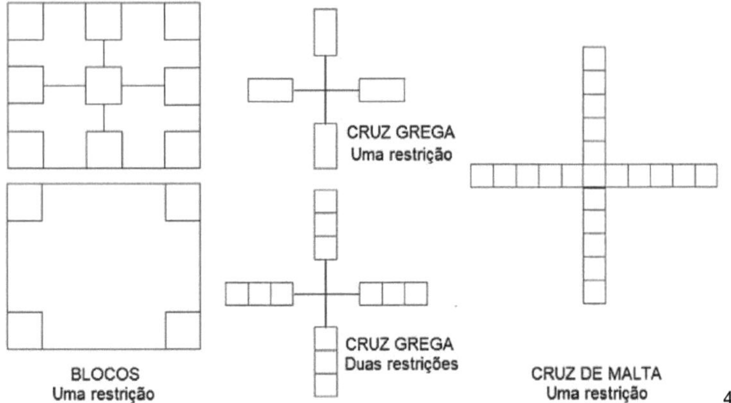

Figura 104 - Examples of conglomerates used in inventories.

CA requires a particular approach to the analysis of sampling variance. The rest of the statistical calculations are similar to SEA.

Analysis of variance in cluster sampling

The analysis of variance is carried out according to the equations for sampling with restrictions. Conglomerate sampling with one restriction is carried out according to Tabela 16.

TABELA 16 - Analysis of variance in cluster sampling with one restriction

Variation Factor	Degrees of Freedom	Sum of Squares	Mean Square	F
Between conglomerates	H - 1	$\sum_{h=1}^{H} n_h \, (\bar{y}_h - \bar{\bar{y}})^2$	SQe / GLe	QMe / QMd
Within the 1st level UAs	n - H	$\sum_{h=1}^{H} \sum_{i=1}^{n_h} (\bar{y}_{hi} - \bar{y}_h)^2$	SQd / GLd	-
Total	n - 1	$\sum_{h=1}^{H} \sum_{i=1}^{n_h} (\bar{y}_{hi} - \bar{\bar{y}})^2$		-

Where: F = Snedecor's F value; H = sample number of conglomerates; n = total number of subunits from all conglomerates ; n_h = number of subunits per conglomerate; $\bar{y}_{hi}$ = average of subunit i in conglomerate h; $\bar{y}_h$ = mean of conglomerate h; $\bar{\bar{y}}$ = overall sampling mean; GL_e = degrees of freedom between conglomerates ; GL_d = degrees of freedom within conglomerates; SQ_e = sum of squares between conglomerates; SQ_d = sum of squares within conglomerates; QM_e = mean square between conglomerates; QM_d = mean square within conglomerates.

If the F value is not significant, there is no significant difference between conglomerates.

Estimates of variance without trend are obtained by (PÉLICO NETTO E BRENA, 1997):

$$s_y^2 = \frac{QM_e + (n_h - 1)\, QM_d}{n_h}$$

$$s_d^2 = QM_d = \frac{\sum_{h=1}^{H} n_h (\bar{y}_h - \bar{\bar{y}})^2}{H - 1},$$

$$s_e^2 = \frac{n_h QM_e - QM_d}{n_h},$$

Intra-cluster correlation coefficient (r)

$$r = s_e{}^2 / (s_e{}^2 + s_d{}^2)$$

Where: r = intra-cluster correlation coefficient; $s_e{}^2$ = variance between clusters; $s_d{}^2$ = variance within clusters.

The closer the intra-cluster correlation coefficient (r) is to zero, the more homogeneous the population is. The population is considered absolutely homogeneous with r=0 and moderately homogeneous with r=0.4.

For cluster sampling, r has an acceptable limit of up to 0.4. If this value is exceeded, there is a high level of heterogeneity in the population and it is necessary to stratify the population into more homogeneous subpopulations.

9.2.15.5 AME - multi-stage sampling

Multi-stage sampling for fauna species is uncommon and we found no examples in the literature, so it is not within the scope of this work.

AME is the sampling process in which the entire population area is divided into large sampling units (AUs), which may not be the same size, called 1st stage AUs; these are redivided into smaller 2nd stage AUs, which can be redivided into even smaller 3rd stage AUs; and so on. A few UAs are then selected at each stage, either randomly or systematically, to make up the pilot inventory sample and determine how many UAs are needed for the desired accuracy.

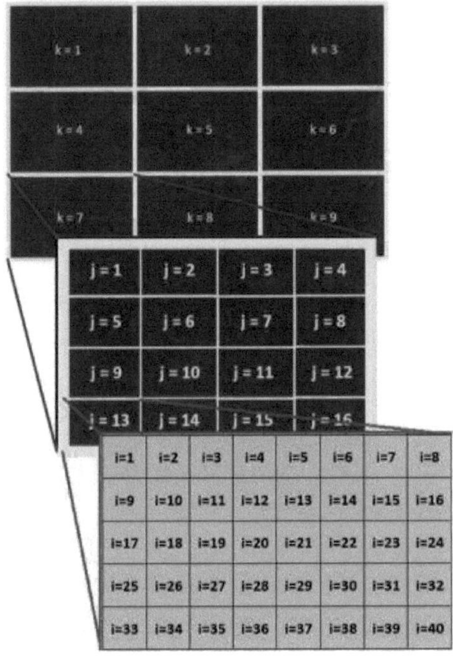

Figura 105 - Multi-stage sampling scheme with two restrictions (1st, 2nd and 3rd stage UAs)

The distribution of AUs in each stage can be either systematic or random. In the pilot inventory, 3 to 10 AUs can be used in each stage, depending on the size and total (potential) number of AUs in each stage.

Multi-stage sampling is a process used for large areas where access to UAs is difficult. The advantage is the reduction in costs and sampling time for the same accuracy compared to other processes.

9.3 Monitoring

Monitoring is a continuous inventory, with occasions repeated periodically in order to detect population variations and trends in the species studied over time.

The appropriate methodology must be chosen for each case. The following must be taken into account when choosing the monitoring methodology:

> Type of data required to achieve the objectives of the inventory and monitoring;

> Spatial extent and duration of the project;

> Life history, biology, ecology and ethology of the species;

> Characteristics of the study area;

> Budget, logistics and time available;

> Familiarity of the observer with the method and species.

9.3.1 Types of monitoring

There are three forms of monitoring:

> Surveillance monitoring;

> Specific monitoring;

> Remote monitoring.

9.3.1.1 Surveillance monitoring

In surveillance monitoring, after distributing the surveillance points, which are the sampling units, surveillance is carried out from each point. At the sampling points, all the animals sighted during pre-established periods are recorded.

The characteristics of surveillance monitoring are as follows:

> There is no prior hypothesis;

> Focus on many species and localities;

> The more information, the better.

This type of monitoring can be carried out from watchtowers, camouflaged tents, or with the aid of photographic or video cameras, triggered by the detection of animal movement.

9.3.1.2 Specific monitoring

The main characteristics of specific monitoring are as follows:
- It is oriented towards a target species or a specific issue;
- It's based on a prior hypothesis;
- The design is adequate and well defined.

9.3.1.3 Remote monitoring

Remote monitoring is carried out at a distance with the main objectives of delimiting the area of occurrence, daily and seasonal migrations to feeding and breeding sites.

Remote monitoring is carried out using resources that make it possible to determine exactly where an animal or group of animals is at the time of measurement, such as:
- Tracking by recapture of ringed and chipped animals;
- Satellites for large animals or groups of medium and large animals;
- Installation of GPS locators on individuals;
- Installation of radio transmitters in individuals.

9.3.2 Monitoring in planted forest areas

"The conservation of wild fauna in forested areas is recognized as being of vital importance in biological stability, the maintenance of biodiversity, the biological control of pests, the maintenance of the aesthetic values of nature and the processes of vegetation renewal in nature reserves." (ALMEIDA and ALMEIDA, 1998)

The main objective of monitoring fauna in areas where production forests are cultivated and managed is to evaluate aspects that help conserve fauna and their habitats in forestry projects aimed at sustainable development.

Ten questions to answer when monitoring fauna in planted forests (ALMEIDA and ALMEIDA, 1998):

"1 - What is the minimum size of a natural vegetation reserve in forested areas?

2 - What percentage of natural areas should be maintained in relation to forested areas?

3 - What is the ideal distribution of nature reserves in forested areas?

4 - Which is better, a single large reserve in a single location, or many small reserves spread out?

5 - What is the minimum viable area required by the various wildlife populations?

6 - What is the minimum viable size of each wild animal population for them to be safely conserved?

7 - Will typical forest species of wildlife be able to be maintained in areas of ecological tension that characterize forest fragments?

8 - Do the forested areas surrounding nature reserves allow for the exchange of genetic material from the species of fauna remaining in the nature reserves held within them?

9 - Is the understory important and necessary for the dispersal of fauna species in forested areas?

10 - Are "biological bridges" or forest corridors really functional in connecting nature reserves, allowing the passage of specimens and the exchange of genetic material, thus reducing inbreeding problems?"

Guilds, edge species, mixed flocks, migrants, rare or endangered species and forest interior species with low dispersal capacity are valuable information for monitoring and managing fauna and their habitats.

A guild is a group of species that exploit the same type of natural resource (e.g. granivore guild; pelagic guild; trophic guild). Species are grouped together when there is a significant overlap of niches, without taking their taxonomic position into account.

9.3.3 Habitat mapping and monitoring

Natural Vegetation Reserves or Wildlife Reserves must be monitored in accordance with the following conditions:

- ➢ Geographical position of the reserve;
- ➢ Type of vegetation;
- ➢ Conservation status of vegetation cover;
- ➢ Vertical heterogeneity;
- ➢ Reserve dimensions;
- ➢ Distance from other reserves.

Monitoring should be carried out with the aid of:

- ➢ Geographical maps;
- ➢ Maps;
- ➢ Aerial photos;
- ➢ Satellite images (false color);
- ➢ Field confirmation is essential;
- ➢ Phytosociological studies of areas of natural vegetation are valuable for assessing the state of the vegetation and the potential for supporting fauna.

9.4 Inventory planning

The recommended sequence of actions for a wildlife inventory is suggested below:

- ➢ Visit the area to be inventoried beforehand;
- ➢ Carry out a literature review on the location and species of occurrence;
- ➢ Map the area and phytophysiognomies;
- ➢ Planning office and field work and compiling data;
- ➢ Train the execution team;
- ➢ Carrying out inventory activities;

- Carry out scientific identification of the species;
- Process the data;
- Drawing up occurrence and distribution maps;
- Draw up the report;
- Present the report to interested parties;
- Review the report and make any changes suggested during the presentation;
- Make corrections and final adjustments;
- Publish the final report.

A fauna inventory report must include at least the following sections:
- Introduction;
- Objectives;
- Methodology;
- Results;
- Conclusions;
- References;
- Appendices and annexes.

The results of the survey and statistics determined for the species and populations of the inventoried area may include:
- List of species and number of specimens detected of each in the sampling;
- Abundance - estimate of the total number of individuals in a population;
- Population density - number of individuals per unit area (specimens/km²);
- Detectability - The probability of a target being observed, heard, or captured in a sampling unit in a given time - the ratio between the number of individuals detected per unit of time (specimens/hour).
- Reproductive rate - annual number of live births per female;
- Fertility rate - the number of children a female would have by the end of her reproductive life;

> Population growth rate - increase in population each year;
> Geographical distribution - area where the species has been found living naturally.

The list of species found should indicate:
> Popular name and scientific name;
> Origin (native or exotic);
> Degree of threat;
> Potential invader;
> Environmental quality indicator;
> Endemism;
> Rarity.

10 CONSERVATION AND PRESERVATION OF WILDLIFE

The purpose of nature conservation and preservation is to protect the environment, but with different approaches. The aim of preservation is to protect the environment from the harmful effects of human activity, while conservation aims to protect the Earth's natural resources for current and future generations (NATIONAL GEOGRAPHIC, 2023a).

Wildlife preservation is the process of maintaining original wildlife conditions without human interference. Wildlife conservation is the practice of protecting plant and animal species and their habitats. As part of the world's ecosystems, wildlife provides balance and stability to nature's processes. The aim of wildlife conservation is to ensure the survival of these species and to educate people about how to live sustainably with other species (NATIONAL GEOGRAPHIC, 2023).

10.1 Conservation and preservation

In Brazilian legislation, Law 9.985/2009 defines:

> ➢ "nature conservation" as: "the management of human use of nature, including the preservation, maintenance, sustainable use, restoration and recovery of the natural environment, so that it can produce the greatest benefit, on a sustainable basis, for present generations, while maintaining its potential to meet the needs and aspirations of future generations, and ensuring the survival of living beings in general";
>
> ➢ and "preservation" as: "a set of methods, procedures and policies aimed at the long-term protection of species, habitats and ecosystems, as well as the maintenance of ecological processes, preventing the simplification of natural systems".

It can therefore be understood that conserving means using natural resources in a sustainable way, guaranteeing their availability for future generations; while preserving means keeping natural resources in their original form, preventing them from being altered by man.

Actions for the sustainability of natural resources aim to maintain wildlife and biological diversity in current conditions so that future generations will find the environment in the same condition as the current generation.

Preservation is necessary when there is a risk of biodiversity loss, whether of a species, an ecosystem or a biome as a whole.

Conservation of wildlife and its biological diversity, represented by genetic resources, can be maintained *in situ, ex situ* and *on farm,* as follows:

> ➢ *IN SITU* CONSERVATION - Work to maintain the integrity and diversity of species within the ecosystems that make up their natural habitat. It is mainly carried out in Conservation Units (UC), Permanent Preservation Areas (APP) and Legal Reserves (RL).
>
> ➢ *EX SITU* CONSERVATION - A conservation effort based on maintaining the original genetic variability outside the natural habitat, such as captive breeding and germplasm banks. It is the maintenance, outside the natural habitat, of representative populations or genes of the target species, of scientific, economic or social importance.
>
> ➢ *ON FARM* CONSERVATION - This is a complementary strategy to *in situ* conservation; it is a process that allows species to continue their evolutionary process and is one of the ways of genetically conserving agrobiodiversity, especially species of economic interest.

10.2 Conservation management

The concept of management is defined according to human influence on an ecological system. Wildlife control can prevent or mitigate the loss of biodiversity. Managing populations for conservation implies knowledge about the population of the species to be managed. The aim of management is to maintain the balance of populations and prevent any species from becoming extinct.

10.2.1 Information needed for proper management

10.2.1.1 On the biology of the species:

- ➢ Eating habits,
- ➢ Behavior and interaction with other species,
- ➢ Population structure,
- ➢ Population dynamics,
- ➢ Birth and death rates,
- ➢ Immigration and
- ➢ Emigration.

10.2.1.2 Ecosystem where the population is found:

- ➢ Environment: soil, climate, vegetation, etc;
- ➢ Other species of the ecosystem's biota.

10.2.1.3 A wild population can be managed to:

- ➢ Make it grow;
- ➢ Make it smaller;
- ➢ Exploit it sustainably;
- ➢ Let it run its course and monitor it.

10.2.2 Conservation strategies for endangered species

The characteristics of Brazil mean that the country has a great responsibility in terms of preserving life on the planet, due to its extension of more than 8 million km², one of the greatest wealth of species in the world that live in six terrestrial biomes and three large marine ecosystems, with more than

100,000 animal species and more than 43,000 species of flora (MMA, 2023). Brazil's different climatic zones have led to the formation of the six terrestrial biogeographical zones (biomes) and marine ecosystems listed below:

> ➢ Amazon rainforest - the largest tropical rainforest in the world;
> ➢ Pantanal - the largest floodplain on the planet;
> ➢ Cerrado - a huge area of savannas and woodlands;
> ➢ Caatinga - a unique biome made up of semi-arid forests;
> ➢ Pampas - large areas of pasture and deciduous riparian forests;
> ➢ Atlantic Rainforest - a large expanse of tropical rainforest bordering the ocean.
> ➢ Marine coast of 3.5 million km² - includes ecosystems such as coral reefs, dunes, mangroves, lagoons, estuaries and swamps.

With a view to protecting Brazil's biological heritage, the Ministry of the Environment (MMA) established the National Program for the Conservation of Endangered Species (Pro-Species) with Ordinance MMA No. 43 of January 31, 2014, with the aim of adopting preventive, conservation, management and administrative actions to minimize the threats and risk of species extinction in Brazil.

The MMA (2023) considers that the main causes of extinction are "the degradation and fragmentation of natural environments, resulting from the opening up of large areas for pasture or conventional agriculture, disorderly extractivism, urban sprawl, expansion of the road network, pollution, forest fires, the formation of lakes for hydroelectric dams and surface mining".

The three strategic instruments of the Pro-Species Project are (MMA, 2014):

> I - **Official National Lists of Endangered Species**, with the purpose of recognizing species threatened with extinction in the national territory, on the continental shelf and in the Brazilian exclusive economic zone, for the purposes of restricting use, prioritizing conservation actions and recovering populations;
>
> II - **National Action Plans for the Conservation of Endangered Species-PAN**, drawn up to define in situ and ex situ actions for the

conservation and recovery of endangered and near-endangered species; and

III - **Databases and information systems** aimed at supporting extinction risk assessments, as well as the process of planning conservation actions, with the identification of the areas of greatest biological importance for endangered species and the areas with the highest incidence of anthropic activities that put their survival at risk.

The strategy of the Pro-Species Project is to adopt preventive, conservation, management and administrative actions to minimize threats and the risk of species extinction through four components:

Component 1 - Incorporating the conservation of endangered species into sectoral policies.

> Objective: To promote measures to reduce threats and strengthen the conservation policy framework for threatened species, integrating species conservation into established public policies.
> Method: Develop strategic actions and policies.
> Expected result: Contain habitat loss and degradation by increasing species protection through sectoral policies, the adoption of territorial plans and mitigation measures.

Component 2 - Combating hunting, fishing, illegal logging and trafficking in wild species.

> Objective: To increase effectiveness in tackling illegal or irregular exploitation of biodiversity.
> Method: Promote measures to develop national capacities to combat environmental crime.
> Expected result: To create measures and initiatives to engage local communities in preventing and combating illegal trafficking in fauna and flora.

Component 3 - Early warning and detection of invasive alien species.

> Objective: To create an Early Warning and Detection System for Invasive Species to prevent and control new biological invasions in Brazil.

> Method: Forming a network of actors and drawing up protocols for alerting and detecting invasive alien species, enabling the creation of a system for wide use.

> Expected result: Network of actors working to quickly detect and prevent the entry of invasive alien species.

Component 4 - Coordination and Communication.

> Objective: To ensure the integration, transparency and dissemination of the stages and results of the project and constant involvement with stakeholders.

> Method: Establish and strengthen a network of partners, increasing the dissemination of information on biodiversity.

> Expected result: To sensitize and engage local communities, states and stakeholders involved in biodiversity for a more active participation in the project components.

MMA Ordinance No. 287, of July 12, 2018, represents a new strategy to prevent the extinction of endangered species with the creation of "Brazilian Alliance for Zero Extinction Sites (BAZE-Sites)", the text of which is reproduced below:

> "ORDINANCE No. 287, OF JULY 12, 2018 - Recognizes the Sites of the Brazilian Alliance for Zero Extinction - Sites-BAZE
>
> THE MINISTER OF STATE FOR THE ENVIRONMENT, in the use of his powers, and taking into account the provisions of Decrees No. 4.339, of August 22, 2002 and No. 8.975, of January 24, 2017, and MMA Ordinance No. 43, of January 31, 2014, and what is contained in Administrative Process No. 02000.000774/2018-01, resolves:
>
> Art. 1 - This Ordinance recognizes the Brazilian Alliance for Zero Extinction Sites - BAZE-Sites - as areas that are home to the last refuges of endangered species, classified in the "Endangered" (EN) or "Critically Endangered" (CR) threat categories, according to the

Official Lists of Endangered Brazilian Fauna and Flora Species, and whose geographic distribution is restricted to one or a few locations very close to each other.

Art. 2 - The BAZE Sites will be used to implement public policies aimed at the conservation and recovery of endangered species, and should be detailed on maps published by means of Ordinances issued by the Minister of State for the Environment and considered for the identification of Priority Areas for the Conservation, Sustainable Use and Benefit Sharing of Brazilian Biodiversity.

Art. 3 - The identification and updating of BAZE Sites, with their respective species, will be carried out whenever the Official National List of Endangered Fauna and Flora is updated.

§Paragraph 1 - The updating of BAZE Sites will follow a specific methodology and will include consultations with specialists and the Technical Chamber of Threatened Species of the National Biodiversity Commission - CONABIO.

§Paragraph 2 - Information on BAZE Sites and their identification process will be available on the Ministry of the Environment's website.

Art. 4 - MMA Ordinance no. 182 of May 22, 2006 is hereby revoked.

Art. 5 - This Order shall enter into force on the date of its publication."

10.2.3 Conservation strategies for endemic species

The conservation of endemic species with little genetic diversity is a complex challenge, but not an impossible one. These species are extremely vulnerable to extinction because they have small populations and generally little genetic variability. They are considered essential to their ecosystems and become a thermometer if we want to measure the state of health of a territory, making their protection essential (IBERDROLA, 2024). There are various strategies that can be adopted to protect these species, such as:

> ➢ Habitat restoration - Restoring and protecting natural habitats is fundamental to the survival of endemic species. This includes creating protected areas and mitigating the threats of habitat degradation.
> ➢ Maintaining viable populations - Carefully planned captive breeding and reintroduction programs can help increase population size and improve genetic diversity.
> ➢ Constant monitoring - Collecting data on endemic populations, including genetic information, is crucial to understanding conservation status and guiding informed decision-making.
> ➢ Controlled introduction of genes - In some cases, the introduction of genes from related populations can increase the genetic diversity and adaptability of species.
> ➢ Education and awareness - Involving local communities and raising awareness about the importance of conserving endemic species is essential to ensure the necessary support.

10.3 National Action Plans NAP

"The National Action Plans for the Conservation of Endangered Species - PAN are management tools, built in a participatory manner, for the planning and prioritization of actions for the conservation of biodiversity and its natural environments, with objectives established over a defined time horizon" (ICMBio, 2024).

There are currently 45 NAPs underway, covering 1,054 threatened species (Tabela 17), which represent 84% of threatened species and on which 3,560 actions have been established, the progress of which is quantified in Figura 106.

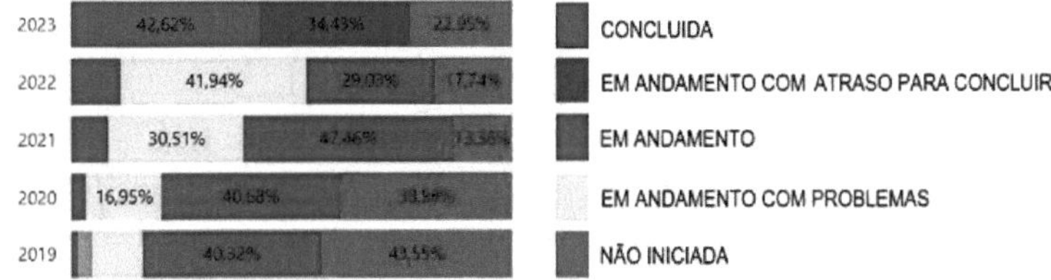

Figura 106 - Progress of PAN actions. Source: ICMBio (2024).

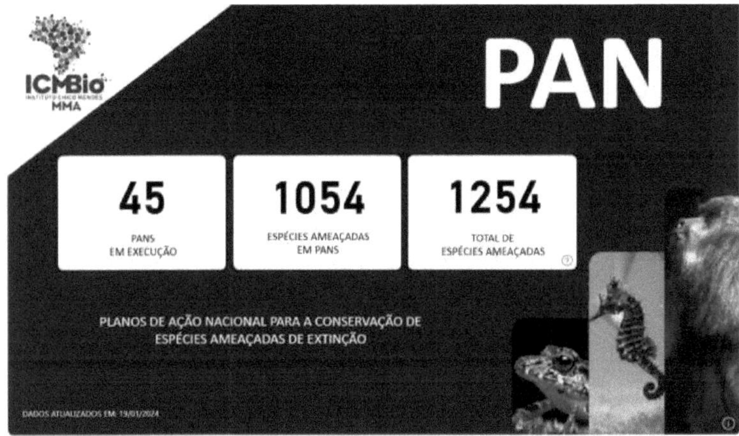

Figura 107 - Number of PANs and threatened species included in PANs. Source: ICMBio (2024).

ICMBio Normative Instruction No. 21, of December 18, 2018, regulates the procedures for drafting, approving, publishing, implementing, monitoring, evaluating and revising National Action Plans for the Conservation of Endangered Species. Each PAN cycle is valid for at least five years.

TABELA 17 - Animals included in PANs by level of threat (ICMBio, 2014)

Animal categories	Threat level		
	CR	EN	VU
Amphibians	23	6	9
Birds	35	79	134
Freshwater invertebrates	21	12	6
Marine invertebrates	4	2	13
Terrestrial invertebrates	72	67	36
Mammals	9	37	43
Aquatic mammals	2	6	4
Continental fish	96	109	76
Marine fish	31	14	36
Reptiles	8	43	20

ICMBio (2018d) provides guidance on the preparation and implementation of PANs in the "Guide for the management of national action plans for the conservation of endangered species".

The process of drawing up and implementing the NAP must take into account:

> "I - the principles of strategic and tactical planning, with a clear definition of the improvement desired in the conservation of the species and environments targeted by the PAN;
>
> II - the involvement of actors who are relevant to reducing threats to biodiversity;
>
> III - definition of the future vision, general objective, specific objectives and actions, demonstrating the causal relationship between them, and focusing on the main threats to be reduced or suppressed;
>
> IV - indication of feasible actions within the timeframe of the plan and within the governance of the actors involved;
>
> V - establishing indicators and targets to verify the achievement of the NAPs' objectives;
>
> VI - transparency and publicity in the preparation, implementation, monitoring, evaluation, review and dissemination of the PAN;
>
> VII - establishment of a continuous monitoring, evaluation and review process; and
>
> VIII - shared search with partner institutions for means to implement PAN actions." (IN ICMBio n° 21/2018)

The PAN Proposal must be drawn up by the CNPC and submitted for technical analysis by COPAN, forwarded for approval by CGCON and must contain:

> "I - conservation targets (endangered species and/or environments);
>
> II - contextualization of threats to conservation targets;
>
> III - justification for the construction of the PAN and conservation opportunities;
>
> IV - schedule of meetings and workshops;
>
> V - estimated costs of meetings and workshops;
>
> VI - the team responsible for drawing up the PAN; and
>
> VII - appointment of the server responsible for coordinating the PAN, designated by the CNPC Coordinator." (IN ICMBio n° 21/2018)

The NAP must also define the geographical scope of action.

A good conservation action plan for an endangered species should include the following chapters and sections:

1. Introduction - history, folklore, conflicts with the species, general situation of the species' population(s), the species' extinction risk category and criteria in regional, national and IUCN assessments, among other general information.

2. Taxonomy of species - common names, phylum, class, order, family, genus and species.

3. Biological and life aspects - size, morphological features, colors, characteristics for identifying the species and specimens; eating habits and preferred foods, social and protective habits, times of greatest activity; mating season and rituals, protection of offspring, size of the area occupied by male and female individuals; period of dependence of offspring on parents; functions of the species in the ecosystem; etc.

4. Geographical distribution - regions of occurrence, distribution map if possible, populations and places of highest and lowest

population density; ecosystems it inhabits; occurrence in protected areas and population status.

5. Natural history of the species;
6. Stakeholders relevant to threat reduction invited to participate;
7. Threats - examples of threats caused by: habitat alteration (reduction, degradation and destruction); mining; trampling on transport routes; conflicts with invasive species; overexploitation by predatory hunting, wildlife trafficking, sport, food and inputs; pollution by fertilizers, pesticides, plastics, medicines, organic waste and metals; the impact of climate change on the species; diseases; among others.
8. Conservation actions already carried out or in progress;
9. Conservation strategies and actions. The following are actions that can be used to protect endangered species:
   - Population and Habitat Viability Analysis (PHVA);
   - In situ conservation - choosing sites for: monitoring, population protection, reintroduction and translocation;
   - Ex situ conservation - captive breeding, translocation of individuals to increase genetic variation in wild populations;
   - Developing captivity and translocation protocols;
   - Study the causes of mortality in captivity in order to reduce it;
   - Promote integration between educational and research institutions and organizations in favor of the species;
   - Environmental education - bringing knowledge and raising awareness through schools, landowners, representatives of society, inspection bodies and the political environment;
   - Study the diseases that affect the species and ways of reducing infections;

- Signpost traffic routes and promote the installation of equipment for animals to cross them;
- Obtain information on the demography of the species;
- Develop research into the ecology of the species to improve protection actions;
- Monitoring the movement of individuals and populations in order to forecast their development and population growth and determine their behavior;
- Plan and implement actions to minimize socio-economic conflicts with the species in its area of occurrence;
- Seek resources for research and conservation of the species;
- Promote the conservation of genetic material from populations of the species, including eggs and semen;
- Promote the proper disposal of seized specimens;
- Publicize the program in the media;
- Make scientific publications about the species;
- Induce the creation of policies and legislation to support the conservation of the species.

10. List of existing closed, ongoing and necessary research.
11. Monitoring, evaluations and revisions of the plan;
12. Expected results;
13. Budget.

11 WILDLIFE MANAGEMENT

The actions of human civilization have led to a wave of extinctions of wild species. The probability of extinction increases with the reduction in animal populations and the extinction of a species is irreversible. The variables that lead to the extinction of reduced populations are not always identified. Species need a minimum number of individuals to survive and evolve.

When the number of individuals of a species is reduced to a critical level, Conservation Management must be adopted to guarantee minimum levels compatible with its perpetuation (VALLADARES-PÁDUA, 2006), including its genetic variability, demographic density and ecological factors.

The main actions to prevent the extinction of a species are to reduce the threats to the species and restore the viability of the species and the habitats in which it occurs.

Threats to the existence of species that can be controlled or reduced may be:

> - Destruction of nature and fragmentation of ecosystems;
> - Habitat degradation (including pollution);
> - Overexploitation of species;
> - Introduction of exotic species;
> - Increased occurrence of diseases.

The challenges for conservation therefore involve reducing threats and genetic and demographic conservation management.

Species can be threatened by factors:

> - Intrinsic - random variations in genetic and demographic events;
> - Extrinsic - environmental events acting on the genetics and demography of the species.

Small populations do not have the same health, productivity and resilience as large populations and the threats to their populations include:

- ➢ Intrinsic factors
  - ○ Demographics affect: birth rate; death rate; sex ratio of the population.
  - ○ Genetic intrinsic factors: random gene loss.
- ➢ Extrinsic factors:
  - ○ Extreme environmental variations;
  - ○ Disasters: fires, cyclones.

Small populations do not have the same range of survival opportunities as larger populations, especially those endemic to small regions, and should be prioritized.

The recovery of reduced populations involves the following aspects:
- ➢ Habitat and species are inseparable;
- ➢ Habitat protection and recovery must go hand in hand with actions to reduce threats to the species and restore its viability;
- ➢ The first step is to diagnose the situation of the species and its habitat, defining the threats to its survival;
- ➢ There are cases where *in situ* actions are sufficient, but in others there is a need for *ex situ* actions such as captive breeding;
- ➢ Small populations require the creation of a scenario with sufficient gene flow to avoid deleterious genetic and demographic effects, increasing the number of individuals and maintaining the genetic variability of the species.

Maintaining gene flow through translocations and reintroductions into small populations aims to reduce the chances of inbreeding. Inbreeding is a factor that reduces the survival capacity of species, especially small isolated populations.

Gene flow can be maintained by:
- ➢ Reintroduction - moving animals born in captivity back into the wild, within their original area of occurrence.
- ➢ Translocation - movement of wild animals between different subpopulations within the area of occurrence.

Reintroduction can be carried out by captive populations under the following conditions:

> They must show great genetic variability between individuals;
> New wild individuals should be added periodically to avoid inbreeding;
> Keeping the animals as wild as possible - avoiding contact between the animals and humans and providing conditions as close to the wild as possible to make the reintroduction as successful as possible;
> Wildlife research provides information for developing captive breeding standards;
> Combining actions in captivity and in the wild is the best way to conserve endangered species.

Translocations from wild populations represent human-induced gene flow and also aim to prevent inbreeding in populations, which can be caused by:

> Isolation of populations on islands, due to fragmentation by human action, can induce wildlife populations to decline and even become locally extinct.
> Larger nature reserves may also lack a viable number of individuals of large species for long-term survival.

Management in these cases requires actions in the wild (*in situ*) and in captivity (*ex situ*).

Isolated populations in fragments, with or without some gene flow between them, are called subpopulations; and the set of them is called a metapopulation.

In Metapopulations, gene flow can be increased through reintroductions and translocations, including a core population bred in captivity.

In metapopulation management, the core population in captivity needs ample genetic diversity, obtained by periodically introducing new wild individuals from different sources. The core population will serve to produce individuals for reintroduction into the subpopulations. Animals in captivity must be kept wild to enable reintroduction to the natural environment.

Translocations of wild individuals between subpopulations are also used to increase gene flow.

To facilitate natural gene flow between subpopulations, biological corridors can be created.

It is important to maintain and restore the quality of the natural habitat.

Movements should follow IUNC protocols whenever possible.

11.1 Metapopulation management and conservation

Metapopulation management involves integrated actions between captivity and the wild, focusing on the conservation of small subpopulations of a species together as a large population. The captive population (core population) does not need to be large, because the constant flow of individuals between captivity and wild populations will maintain the necessary genetic diversity, with a smaller number of animals, as long as the origin of the captive animals is diverse and the replacement of individuals is frequent (OLIVEIRA, 1012).

The core population will always have a high proportion of wild gene diversity. This makes it a population with the same genetic and demographic quality as the wild population, and it can be used to repopulate nature in the event of adverse factors.

In metapopulation management, maintaining population viability is not limited to the captive population, as there is integration between the wild subpopulations and the core population, with the main emphasis being on the wild subpopulations, which represent the basis for the species' viability (OLIVEIRA, 2012).

Metapopulations require ongoing management (IUCN 2014-SSC), with some translocations requiring management over many years; monitoring their results ensures a basis for continuing or changing management regimes (Figura 109). It also provides a justification for any change in the translocation objectives or schedule. Experience from the results of a translocation makes it possible to improve the process using a more formal adaptive management approach, in which alternative models are defined and tested during monitoring.

For translocations, a risk analysis should be carried out, including the evaluation of options that reduce the risk of undesired results. The preferred course of action in the event of unwanted results is to remove the translocated population, however this is usually only possible in the early stages after establishment, when unwanted effects may not yet be evident.

Conservation translocations (Figura 108) consist of reinforcement and reintroduction within the native range of a species and introduction for conservation purposes, comprising assisted colonization and ecological replacement, outside its native range (IUCN-SSC, 2014).

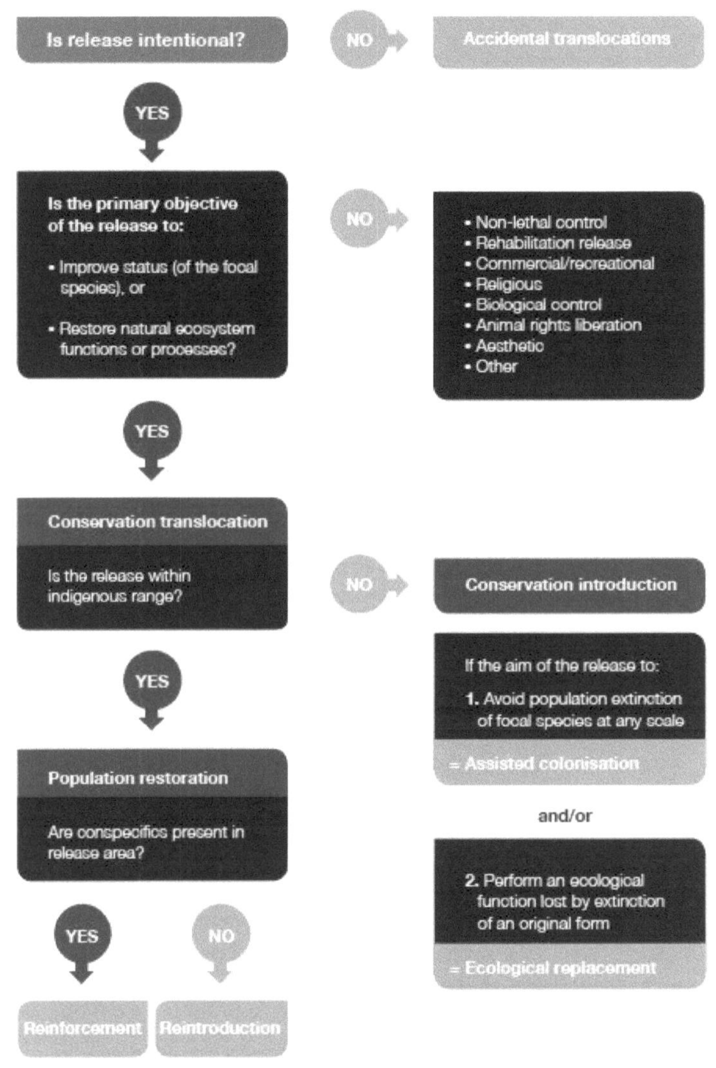

Figura 108 - Scheme for translocations. Source: IUCN-SSC (2014).

In the ongoing management of metapopulations, the IUCN-SSC (2014) recommends:

> 1) Monitoring the information allows researchers to assess whether the objectives are being achieved according to schedule. This

information can then be used to adjust any ongoing management of the current population and, more broadly, to help plan further translocations.

2. Adjustments can involve increasing or decreasing the intensity of management or changes in the type of management. For example, if a translocated population fails to grow despite ongoing management, it may make sense to increase the intensity of that management. Alternatively, it may be better to try a different management option or even cancel the management and relocate the remaining individuals to another area. If monitoring indicates that the translocated population is causing unwanted impacts, this could potentially lead to a decision to control or remove the population and conduct other management actions to mitigate these impacts. The decision-making process should be transparent, and reflect the current understanding of population dynamics and impacts, the value placed on different possible outcomes by all those involved, and the costs of management options.

3. While decisions have to be made, it is essential to consider the uncertainty in population predictions. There are two sources of uncertainty in these predictions. First, populations are subject to random variations due to the role of chance in the fate of each individual (demographic stochasticity) or environmental fluctuations (environmental stochasticity). Second, understanding of populations is always limited, and decisions must be supported by including the assumptions behind them and the extent of uncertainty in biological knowledge of them.

4. A key benefit of monitoring is that it allows researchers to progressively improve understanding and then develop more accurate models for subsequent predictions and target setting. This is especially useful when the original objectives cannot be met due to factors beyond management control. This process of learning from the results is called "adaptive management". However, adaptive management does not merely mean adjusting management after monitoring; it means having clearer models that are then evaluated against the monitoring results. Sometimes it is appropriate to deliberately manipulate management options in order to accumulate knowledge, which is a process known as "active adaptive management". For example, if a translocated population is growing towards the goal set in a given management regime, it may make sense to temporarily cancel the regime to ensure that it is really necessary."

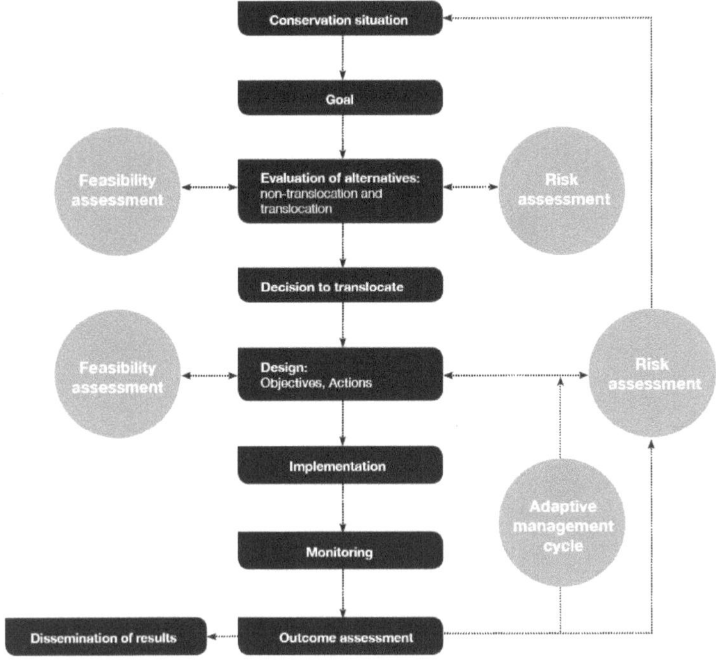

Figura 109 - Translocation flowchart for conservation purposes. Source: IUCN-SSC (2014).

In addition to habitat conservation, wildlife corridors and improving gene flow between subpopulations, other actions must be taken to ensure the success of a wildlife management and conservation program, such as obtaining resources, environmental education and political action.

Metapopulations can have four different types of structure, as in the diagrams in Figura 110 below.

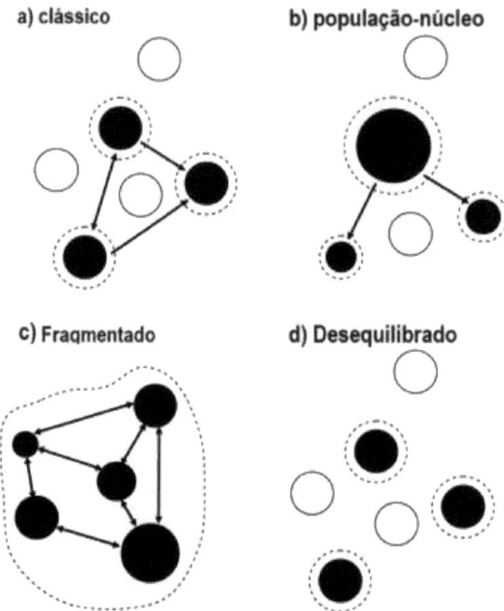

Figura 110 - Conceptual diagrams of metapopulation dynamics for four spatially diverse populations: (a) classic metapopulation, (b) metapopulation with core population, (c) fragmented population and (d) metapopulation out of equilibrium. Filled circles represent occupied habitat areas; empty circles, empty habitat patches; dotted lines, local population boundaries; arrows, dispersal between patches. Source: Collinge (2009).

11.2 Integrated Conservation Management

Integrated management for the conservation of wild animal species consists of (Valladares-Pádua, 2006.):

> ➤ Activities in nature;
>
> ➤ Activities in captivity;
>
> ➤ Complementary activities.

11.2.1 Activities in nature

Survey - current status of the species and its area of occurrence; diagnosis of threats to the species and its habitat;

214

Census and demography - phytosociology, target population size by sex and life stage; diagnosis of the habitat's carrying capacity;

Genetic analyses - studies, by subpopulation, of sexing, polymorphism, karyotype, heterozygosity and molecular genetics are important to help choose animals for transposition or captive breeding;

Ecology and behavior - long-term monitoring of animals: individual identity, age, groups, reproduction, quality of territory, feeding, use of space, times by activities and sites visited.

Habitat reconstruction - restoration and rehabilitation of vegetation to sustain the target population, including with local partnerships, creation of biological corridors.

11.2.2 Captive activities

Basic colony data - identification of individuals, date of birth or entry, date of death or departure, affiliation, place of origin, medical records (health and behavior);

Quantitative genetic analysis - pedigree registration, identity of colony founders, participation of founders, identification of genetic bottlenecks, registration of individual offspring, determination of inbreeding coefficients;

Biochemical genetic analysis - karyotyping, polymorphism and heterozygosity, satellite DNA, nuclear and mitochondrial DNA;

Demographic analysis - current population size, captivity carrying capacity, age and sex structure, age curves for fecundity and survival, changes in self-sustainability, population stabilization for self-sustainability, reproduction and behavior records,

Breeding standards - creating a guide to managing the species in captivity.

11.2.3 Complementary activities

> Environmental education -'schools, neighboring communities, landowners in the area of occurrence, environmental officials;

> Movements - depends on the species, objectives and situation; there are IUCN protocols for various cases;

> Management Committee - a multidisciplinary group of people who draw up and guide the implementation of the program - the committees are usually coordinated by IBAMA;

> Program formulation - clear recommendations for all stages;

> Implementing the program - publicizing it to everyone involved, training the teams, dividing up tasks between teams;

> Reviews and modifications - periodic reviews should be carried out according to the criticality of each activity, including preventive and corrective measures, and new additions for improvement;

> Resources and means - institutions that breed species in captivity, government agencies, etc;

> Influence on public policy - a conservation program involves contact with many sectors of society that can mobilize to influence the creation of rules and legislation pertinent to the management and conservation of wild animal species;

> Expected results and future scenarios - Special software makes it possible to predict results in terms of the future genetics of populations; some habitat and population viability analysis models help predict future survival conditions; creating different scenarios helps choose the best course of action.

11.2.4 Flow of actions in a wildlife conservation and management program

The flow of actions in a wildlife conservation and management program is listed below:

1) Research;
2) Search for resources for the program;

3) Drawing up the plan;
4) Environmental education;
5) Political actions;
6) Program execution;
7) Recording results;
8) Periodic evaluation of the program;
9) Reviewing and improving the program.

11.2.4.1 Research

Literature review on:

- ➢ Environment;

- ➢ Fauna;

- ➢ Flora.

Research in nature:

- ➢ Phytosociological inventory;

- ➢ Inventory and diagnosis of fauna.

11.2.4.2 Searching for resources

- ➢ Government tenders;

- ➢ Funds for nature;

- ➢ Non-Governmental Organizations - NGOs;

- ➢ Civil Society Organizations of Public Interest - OCIP;

- ➢ Environmental Conduct Adjustment Terms - TAC.

11.2.4.3 Drawing up the plan

This is the phase in which the plan is drawn up. The plan must be peer-reviewed. If necessary, it can be presented at a public hearing.

11.2.4.4 Environmental education

- ➢ Neighboring communities;

- ➢ Rural landowners in the area;

- ➢ Government authorities;

- ➢ Community schools;

- ➢ Events on environmental conservation.

11.2.4.5 Public actions

> ➢ Influence on public policies (creation of norms and laws);
>
> ➢ Wildlife conservation events;
>
> ➢ Partnerships with society in general.

11.2.4.6 Program execution

During this phase, all the actions set out in the plan are carried out. Implementation must be accompanied by extensive publicity and environmental education activities. Whenever possible, local communities should be involved in implementation. Reintroduced or translocated animals should be permanently monitored.

11.2.4.7 Recording results

All actions should be recorded, as well as their immediate and medium- and long-term results, so that analysis and corrections can be made. Preferably, the results should be recorded in previously prepared spreadsheets.

11.2.4.8 Periodic evaluation of the program

The team responsible should carry out a periodic evaluation and issue a report. The results should be disseminated to all those involved and included in environmental education actions.

11.2.4.9 Reviewing and improving the program

Actions whose results are insufficient should be analyzed in order to implement process improvements. Planning should be reviewed and all necessary improvements incorporated into it.

12 REFERENCES

ALENCAR, K. **What was marine life like before the Mariana disaster?** UFC Agency, 28/08/2018. Available at: https://agencia.ufc.br/como-era-a-vida-marinha-antes-do-desastre-em-mariana/. Accessed on: 2024.

ALMEIDA, A. F.; ALMEIDA, A. **Monitoring fauna and their habitats in forested areas**. Série Técnica IPEF, v.12, n.31, p.85-92, 1998.

ALMEIDA, J. P. **The extinction of the rainbow: ecology and history**. Rio de Janeiro: Edelstein Center for Social Research, 2008. ISBN 978-85-99662-69-4. Available at: <https://static.scielo.org/scielobooks/s4jcj/pdf/almeida-9788599662694.pdf>. Accessed on: 2023.

AUR, D. List of animal ingredients in products you can't even imagine. **GreenMe**, 06/04/2021. Available at: <https://www.greenme.com.br/consumir/consumo-consciente/80171-produtos-contem-ingredientes-origem-animal/>. Accessed on: 03/04/2023.

AZEVEDO, A. L. Yellow fever: Fiocruz vaccine successfully tested on monkeys. **O Globo/Medicina**, 23/04/2023. Available at: <>. Accessed on: 2024.

BATISTA, C. **Food chain**. All Matter, 2011. Available at: <https://www.todamateria.com.br/cadeia-alimentar/>. Accessed on: 04/04/2023.

BELCHIOR, D. C. V.; SARAIVA, A. S.; LÓPES, A. M. C.; SHEIDT, G.N. impacts of pesticides on the environment and human health. **Cadernos de Ciência & Tecnologia**, Brasília, v. 34, n. 1, p. 135-151, jan./abr. 2014.

BILA, D. M.; DEZOTTI, M. Fármacos no Meio Ambiente. **Química Nova**, Rio de Janeiro, v. 26, n. 4, p. 523-530, 03/02/2003. Available at: <http://www.scielo.br/pdf/qn/v26n4/16435>. Accessed on: 2024.

BLOG ERA DA ÁGUA. Acid Mine Drainage - characteristics and environmental impacts. 12/01/2022. Available at: <https://www.hidroplan.com.br/site/blog-era-da-agua/52-drenagem-acida-de-mina-caracteristicas-e-impactos-ambientais>. Accessed on: 2024.

BRAHNEY, J.; HALLERUD, M.; HEIM, E.; HAHNENBERGER, M.; SUKUMARAN, S. Plastic rain in protected areas of the United States. Science, v. 368, n. 6496, p. 1257-1260, 12/06/2020.

BRAZIL. Decree No. 23.672, of January 22, 1934.

BRAZIL SCHOOL. Available at: https://brasilescola.uol.com.br/. Accessed on: 2023.

BUSINARI, M. **Starling: Bird that lives in flocks arrives in Brazil and causes concern**. UOL, 16/12/2021. Available at: <https://noticias.uol.com.br/cotidiano/ultimas-noticias/2021/12/08/ambiente-economia-e-saude-em-risco-estorninho-se-propaga-no-sul-do-brasil.htm?cmpid=copiaecola>. Accessed on: 2023.

CARNEIRO, J. D. Yellow fever could accelerate extinction of endangered monkeys. **BBC**, 29/05/2017. Available at: <https://www.bbc.com/portuguese/brasil-40024332>. Accessed on: 2024.

CASSINI, S. T. **Ecology - fundamental concepts**. Vitória: PPGEA/UFES, 2005.

CASTILHO, R. **Climate change: what it is, summary, causes and consequences**. Enciclopédia Significados. Available at: <https://www.significados.com.br/mudancas-climaticas-o-que-sao-resumo-causas-e-consequencias/>. Accessed on: 2024.

CASTRO, A. 40 thousand tons a month: The path of organic waste between disposal in the Capital and the landfill in Minas do Leão. **Sul21**, Porto Alegre, 03/02/2021. Available at: <https://sul21.com.br/especiais/40-mil-toneladas-por-mes-o-caminho-do-lixo-organico-entre-o-descarte-na-capital-e-o-aterro-em-minas-do-leao/>. Accessed on: 2024.

CECIERJ. Biology - Fascículo 7 Unidade 17 - Interações Ecológicas - A Teia da Vida. 09/12/2016.

CIVITEREZA, G. **The impacts of mineral fertilization on the environment**. Machado (MG): Terra de Cultivo, 20/05/2021. Available at: <https://www.terradecultivo.com.br/os-impactos-da-adubacao-mineral-no-meio-ambiente/>. Accessed on: 2023.

CRMV-SP. Wildlife management and the duties of the RT. São Paulo, 2023. Available at: <https://crmvsp.gov.br/manejo-de-fauna-e-as-atribuicoes-do-rt/>. Accessed on: 2023.

DGAV. Diseases of Wild Animals. Lisbon: Direção Geral de Alimentação e Veterinária, Portuguese Republic. Available at: <https://www.dgav.pt/animais/conteudo/animais-selvagens/saude-animal/doencas-dos-animais-selvagens/>. Accessed on: 2024.

COELHO, H. A.; CORRÊA, A. A. (coord). **Environmental quality report: Brazil 2020**. Brasília, DF: IBAMA, 2022. Available at: <http://www.ibama.gov.br/sophia/index.php?codigo_sophia=139603>. Accessed on: 2024.

COLLINGE, S. K. **Ecology of Fragmented Landscapes**. Baltimore: Johns Hopkins, 2009.

DNIT. **DNIT studies and monitors wildlife collisions on more than 5,500km of federal highways**. 2014.

ECOA. **Extreme weather events in South America**. ecologiaeacao, 21/12/2023.

ERIKSEN, M.; COWGER, W.; ERDLE, L. M.; COFFIN, S.; VILLARRUBIA-GÓMEZ, P.; MOORE, C. J.; CARPENTER, E. J.; DAY, R. H.; THIEL, M.; WILCOX, C. A growing plastic smog, now estimated to be over 170 trillion plastic particles afloat in the world's oceans - Urgent solutions required. **PLoS ONE** 18(3): e0281596, 8/03/2023. Available at: <https://doi.org/10.1371/journal.pone.0281596>. Accessed on: 2023.

FLORIANO, E. P. **Forest Phytosociology**. São Gabriel: ed. by the author, 2014.

G1 Mato Grosso. Dump will have to be closed due to air risk from birds and planes in MT. G1, Mato Grosso, 20/10/2012. Available at: <https://g1.globo.com/mato-grosso/noticia/2012/10/lixao-tera-que-ser-fechado-devido-risco-aereo-com-aves-e-avioes-em-mt.html>. Accessed on: 2024.

G1 NATURE. **Whale found dead with 40 kilos of plastic in its stomach**. 18/03/2019. Available at: <https://g1.globo.com/natureza/noticia/2019/03/18/baleia-e-encontrada-morta-com-40-quilos-de-plastico-no-estomago.ghtml>. Accessed on: 2024.

GASINE, G. **Plastic, waste & me: unpacked!** Rio de Janeiro: Heirich Böll Foundation, 2022.

GEO-RISKS. Hurricane Katrina. Available at: <https://geo-riscos.no.comunidades.net/furacao-katrin>. Accessed on: 2024.

GRAIPEL, M. E.; CHEREM, J. J.; BOGONI, J. A.; PIRES, J. S. R. Characteristics associated with extinction risk in terrestrial mammals of the Atlantic Forest. **Oecologia Australis**, 20(1): 81-108, 2016

GUIMARÃES, T. The main cause of death of wild animals in Brazil. **BBC**, 02/10/2015. Available at: <https://www.bbc.com/portuguese/noticias/2015/10/150924_atropelamentos_fauna_tg>. Accessed on: 2023.

HALL, S. Volcanoes were key to dinosaur extinction, says new study. **National Geographic**, 09/02/2018. Available at: <https://www.nationalgeographicbrasil.com/animals/2018/02/volcanoes-were-determinant-for-extinction-of-dinosaurs-says-new-study>. Accessed on: 2024.

HYPENESS. **Sperm whale found dead with 29 kilos of plastic in its stomach**. 11/04/2018. Available at: <https://www.hypeness.com.br/2018/04/cachalote-e-encontrado-morto-com-29-quilos-de-plastico-no-estomago/>. Accessed on: 2024.

IBAMA. **Ibama confirms starling invasion in Brazil**. Brasília: Ibama, 16/12/2021. Available at: <https://www.gov.br/ibama/pt-br/assuntos/noticias/2021/ibama-confirma-invasao-de-estorninhos-no-brasil>. Accessed on: 2023.

IBAMA. **Invasive alien species: About invasive alien species**. Brazilian Institute of the Environment and Renewable Natural Resources, 29/11/2022a. Available at: https://www.gov.br/ibama/pt-br/assuntos/biodiversidade/especies-exoticas-invasoras/>. Accessed in 2023.

IBAMA. **Invasive alien species: About the sun coral**. Brazilian Institute of the Environment and Renewable Natural Resources, 29/11/2022b. Available at: <https://www.gov.br/ibama/pt-br/assuntos/biodiversidade/especies-exoticas-invasoras/sobre-o-coral-sol>. Accessed on: 2023.

IBAMA. **Invasive alien species: About the golden mussel**. Brazilian Institute of the Environment and Renewable Natural Resources, 29/11/2022c. Available at: <https://www.gov.br/ibama/pt-br/assuntos/biodiversidade/especies-exoticas-invasoras/sobre-o-mexilhao-dourado>. Accessed on: 2023.

IBAMA. **Invasive alien species: Management and control of wild boar**. Brazilian Institute of the Environment and Renewable Natural Resources, 30/10/2023. Available at: <https://www.gov.br/ibama/en-br/assuntos/biodiversidade/especies-exoticas-invasoras/manejo-e-controle-do-javali>. Accessed on: 2023.

IBDF. Ordinance No. 303, of May 29, 1968.

IBERDROLA. Sustainability: Endemic species, 2024. Available at: <https://www.iberdrola.com/sustainability/endemic-species>. accessed: 2024.

IBGE. Ecosystem accounts: endangered species in brazil, update 2022. IBGE/Environment Coordination, May/2023.

ICMBio. Diagnosis of the invasion of the sun coral (*Tubastraea* spp.) in Brazil. 2018a.

ICMBio. Red Book of Endangered Brazilian Fauna, 1st ed. Brasília, DF: ICMBio, 2018b.

ICMBio. Guide for the management of national action plans for the conservation of **ICMBio** species. Red Book of Endangered Brazilian Fauna, 1st ed. Brasília, v.1, 2018c.

threatened with extinction: PAN - elaborate - monitor - evaluate / ICMBio. Brasília, 2018d. Available at: <https://ava.icmbio.gov.br/pluginfile.php/4592/mod_data/content/20615/PAN%20-%20elabor e%20-%20monitore%20-%20avalie%20%282018%29-v2.pdf>. Accessed on: 2024.

ICMBio. Invasive Alien Species: Information poster on the lionfish for fishermen. Available at: <https://www.icmbio.gov.br/cbc/publicacoes.html#>. Accessed on: 2023a.

ICMBio. Invasive Alien Species: Strategic Guide for Research, Management and Environmental Interpretation Activities on the Lionfish. Available at: <https://www.icmbio.gov.br/cbc/publicacoes.html#>. Accessed on: 2023b.

ICMBio. National Action Plans for the Conservation of Endangered Species - PAN. Available at: <https://www.gov.br/icmbio/pt-br/assuntos/biodiversidade/pan>. Accessed on: 2024.

IUCN-SPC. Guidelines for Using the IUCN Red List Categories and Criteria. Version 15.1. Prepared by the Standards and Petitions Committee, 2022. Available at: <https://www.iucnredlist.org/documents/RedListGuidelines.pdf>. Accessed on: 2023.

IUCN-SSC. IUCN Red List categories and criteria, version 3.1, second edition. Gland, Switzerland and Cambridge, UK: 2012. Available at: <https://portals.iucn.org/library/sites/library/files/documents/RL-2001-001-2nd.pdf >. Accessed on: 2023.

IUCN-SSC. Guidelines for Reintroductions and other Translocations for Conservation Purposes. Gland, Switzerland: IUCN Species Survival Commission, June 2014. Available at: <https://portals.iucn.org/library/sites/library/files/documents/2013-009-pt.pdf>. Accessed on: 2024.

JAMBECK, J. R.; GEYER, R.; WILCOX, C.; SIEGLER, T. R.; PERRYMAN, MI.; ANDRADY, A.; NARAYAN, R.; LAW, K. L. Plastic waste inputs from land into the ocean. **Science**, v. 347, n. 6223, p. 768-771, 13/01/2015.

JOEL, L. Ancient 'Snowball Earth' thawed out in a flash. **Science**, 2 APR 2019. Available at: <https://www.science.org/content/article/ancient-snowball-earth-thawed-out-flash>. Accessed on: 2024.

JOLY, C.A.; SCARANO, F.R.; SEIXAS, C.S.; METZGER, J.P.; OMETTO, J.P.; BUSTAMANTE, M.M.C.; PADGURSCHI, M.C.G.; PIRES, A.P.F.; CASTRO, P.F.D.; GADDA, T.; TOLEDO, P. (eds.). **1st Brazilian Diagnosis of Biodiversity and Ecosystem Services**. Editora Cubo, São Carlos, pp.351, 2019. Available at: <https://doi.org/10.4322/978-85-60064-88-5>. Accessed on: 2023.

KAFRUNI, S; SOUZA, R. Road accidents take the lives of millions of animals a year in Brazil. **Correio Brasiliense**, 09/08/2020.

KAWAI, B.; URIAS, C.; LEONEL, L.; ALMADO, M. Environmental pollution by metals. Available at: <https://www.fernandosantiago.com.br/met90.htm>. Accessed on: 2024.

LEITE, M. B. A. Pollution in the Seas. **Ambiente Brasil**. Available at: <https://ambientes.ambientebrasil.com.br/agua/artigos_agua_salgada/poluic ao_nos_mares.html>. Accessed on: 2024.

MELO, I. S. **Genetic resources**. Brasília: Embrapa, 2021. Available at: <https://www.embrapa.br/agencia-de-informacao-tecnologica/tematicas/agricultura-e-meio-ambiente/manejo/recursos-geneticos>. Accessed in: 2023.

MENUZZI, N.; FACCIN, Y.; PITOL, N.; REIS, M.; KONZEN, M.; DIAS, M. Wildlife trafficking threatens the biodiversity of Brazilian fauna: Illegal trade is responsible for the removal of 38 million animals from Brazil each year. **Revista Arco**, UFSM, 05/09/2020. Available at: <https://ufsm.br/r-601-6260>. Accessed on: 2023.

MESQUITA, J. L. Oil spills at sea, ranking of the worst accidents. **Mar Sem Fim**, Estadão, São Paulo, November 27, 2019. Available at: <https://marsemfim.com.br/ derrame-de-petroleo-no-mar-um-ranking-dos-piores/>. Accessed on: 2023.

METZGER, J. P. WHAT IS LANDSCAPE ECOLOGY? **Biota Neotrópica**, v.1, n.1, 28/11/2001.

MMA. Red book of endangered Brazilian fauna, 1st ed. Brasília, DF : MMA, 2008.

MMA. National Strategy for Invasive Alien Species. Ministry of the Environment, 2019.

MMA. MMA Ordinance No. 43, of January 31, 2014, Institutes the National Program for the Conservation of Endangered Species (Pro-Species). Available at: <http://www.ibama.gov.br/ sophia/cnia/legislacao/MMA/PT0043-310114.pdf>. Accessed on: 2023a.

MMA. Species Conservation. Available at: <https://antigo.mma.gov.br/assuntos-internacionais/blocos/item/15011-conserva%C3%A7%C3%A3o-de-esp%C3%A9cies.html>. Accessed in 2023.

MOSQUERA, P. **Eruption of the Cumbre Vieja volcano in Spain ends after 85 days**. CNN, 25/12/2021. Available at: <https://www.cnnbrasil.com.br/internacional/erupcao-do-vulcao-cumbre-vieja-na-espanha-termina-apos-85-dias-dizem-autoridades/>. Accessed on: 2024.

MPF. **Samarco case**. Available at: <https://www.mpf.mp.br/grandes-casos/caso-samarco>. Accessed on: 2024.

MUNRO, C.; VUE, Z.; BEHRINGER, R.R.; DUNN, C.W. Morphology and development of the Portuguese man of war, *Physalia physalis*. **Scientific Reports**, 9:15522, 2019. Available at: <https://doi.org/10.1038/s41598-019-51842-1>. Accessed on: 02/04/2023.

NATIONAL GEOGRAPHIC. **Conservation**. Available at: <https://education.nationalgeographic.
org/resource/conservation-encyclopedic/>. Accessed on: 2023.

NATIONAL GEOGRAPHIC. **Wildlife Conservation**. Available at: <https://education.national
geographic.org/resource/wildlife-conservation/>. Accessed on: 2023b.

NEVES, C. V.; GAYLARDE, C. C.; BAPTISTA NETO, J. A.; ALMEIDA, M. P.; POMPERMAYER, F. C. L.; VIEIRA, K. S.; FONSECA, E. M. Plastic (and microplastic) in the environment: impacts and challenges for its management. In: PEREIRA, C.; FRICKE, K. (coord.). *IN*: INTERSECTORAL COOPERATION AND INNOVATION: TOOLS FOR SUSTAINABLE SOLID WASTE MANAGEMENT. Braunschweig: Technische Universität Braunschweig, 2022.

NOVAES, J. Environment: concepts, dialogues and reflections - Wildlife Conservation. **ZooSBC**, 2019.

OLIVEIRA, W. S. **Identification of parasitic diseases in endangered fish species in ex situ environments**. Pirassununga - SP: ICMBio, NATIONAL CENTER FOR RESEARCH AND CONSERVATION OF CONTINENTAL FISH, 2012.

PADUA, S. After all, what's the difference between conservation and preservation? **O ECO**, Rio de Janeiro, February 2, 2006.

PARANÁ. Red list of endangered plants in the state of Paraná. Curitiba: State Secretariat for the Environment / GTZ, 1995.

PARANÁ. African snail (Achatina fulica). Curitiba: Paraná State Health Department, 23/06/2021. Available at: <https://www.saude.pr.gov.br/Pagina/Caramujo-Africano-Achatina-fulica>. Accessed on: 2023.

PARRON, L. M. et al (technical editors). **Environmental services in agricultural and forestry systems of the Atlantic Forest Biome** [electronic resource]. Brasília, DF : Embrapa, 2015. Available at: <https://www.alice.cnptia.embrapa.br/alice/bitstream/doc/1024082/1/LivroSer
vicosAmbient
aisEmbrapa.pdf>. Accessed on: 2023.

PEARCE, D.; MORAN, D. **The economic value of Biodiversity**. London: IUCN/Earthscan, 1994.

PEDRO, A. F. P. Death sails on plastic. **Ambiente Legal**, 13/07/2016. Available at: <https://www.ambientelegal.com.br/a-morte-navega-no-plastico/>. Accessed on: 2024.

PÉLLICO NETTO, S.; BRENA, D. A. **Inventário Florestal**. Curitiba, 1997. 316 p.

PORTO, M. L.; MENEGAT, R. Landscape ecology: a new approach to the management of Earth and human systems. In: MENEGAT, R.; ALMEIDA, G. (ORGANIZ.). Sustainable development and environmental management of cities. Edufrgs, 2015.

POULIN, R. **Evolutionary ecology of parasites,** 2 ed. Princeton: Princeton University Press, 2007.

REID W. V.; MILLER K. R. **Keeping options alive**: the scientific basis for conserving biodiversity. Washington DC: World Resources Institute, 1989.

RODRIGUES, C.; FIORENZA, M.; LEMOS COSTA, A.; ERICHSEN, R.; MIRANDA PINTO, J. Raising awareness about animal preservation: a high school experience. ANAIS DO SALÃO INTERNACIONAL DE ENSINO, PESQUISA E EXTENSÃO, v. 8, n. 1, 14 , feb. 2020.

ROSA, C. A.; FERNANDES-FERREIRA, H.; ALVES, R.R.N. Management of wild boar (*Sus Scrofa* Linnaeus 1758) in Brazil. **Biodiversidade Brasileira**, 8(2): 267-284, 2018.

SALAFSKY, N.; SALZER, D.; STATTERSFIELD, A. J.; HILTON-TAYLOR, C.; NEUGARTEN, R.; BUTCHART, S. H. M.; COLLEN, B.; COX, N.; MASTER, L. L.; O'CONNOR, S.; WILKIE; D. A Standard Lexicon for Biodiversity Conservation: Unified Classifications of Threats and Actions. **Conservation Biology**, n.22, p.897-911, 2008.

SANCHES, M. FORESIA - The ride of biology. Insetoland, March 6, 2020. Available at: <https://twitter.com/insetoland/status/1235948074377326592>. Accessed on: 05/04/2023.

SANTOS, V. S. Inquilinismo. **Mundo Educação**. Available at: <https://mundoeducacao.uol. com.br/biologia/inquilinismo.htm>. Accessed on: 2023.

SCHRÖTER, M. et al. Indicators for relational values of nature's contributions to good quality of life: the IPBES approach for Europe and Central Asia. **Ecosystems and People**, v. 16, n. 1, p. 50-69, 2020.

SENA, . T. B. da S. .; SANTANA, . R. C. Animal abuse and the ineffectiveness of legislation. **Revista A Fortiori**, [S. l.], v. 1, n. 1, 2021. Available at: http://revistas.famp.edu.br/revistaafortiori/article/view/296. Accessed on: March 27, 2023.

SIDDIKA, A.; ROBIN, T. I.; USHA, U. R. Constructing a Fully Ranked phylogenetic Constraint Tree using Monophyletic Group, Crown Group and Relative Age Constraints. **URSET**, v.2, n.5, p.236-242, September-October-2016.

SILVA, J. C. R. Zoonoses and Emerging Diseases Transmitted by Wild Animals. **ABRAVAS**, 2004. Available at: <https://bichosonline.vet.br/wp-content/uploads/2016/06/ Ramos-Silva-JC-2004-Doencas-Emergentes-e-Zoonoses-Animais-Silvestres-www-abravas-org-br-.pdf>. Accessed in 2024.

SILVEIRA, L. F. What are fauna inventories for? **Institute for Advanced Studies**, University of São Paulo, v.24, n.68, 2010.

SOUZA, A. K. R.; MORASSUTI, C. Y.; DEUS, W. B. ENVIRONMENTAL POLLUTION BY HEAVY METALS AND THE USE OF PLANTS AS BIOINDICATORS. **Acta Biomedica Brasiliensia**, v.9, n.3, December/2018. Available at: <https://dialnet.unirioja.es/descarga/ articulo/6789234.pdf>. Accessed on: 2024.

SOUZA, J. The number of birds found dead in a bird flu outbreak at the Taim Ecological Station reaches 101. **TERRA**, 27/06/2023. Available at: <https://www.terra.com.br/noticias/chega-a-101-o-numero-de-aves-encontradas-mortas-em-foco-de-gripe-aviaria-na-estacao-ecologica-do-taim,fcbd0c9a221b4196b1196e3bb3a68f37dre1s8xn. html?utm_source=clipboard>. Accessed on: 2024.

SILVA, E. N.; SANTOS, R. S. Mutualism, symbiosis and associated protocooperation between ants, plants and other insects. **Educte**, Maceió, v. 13, n. 1, p. 1943- 1956, 2022. ISSN 2238-9849

SOARES FILHO, B. S. **Análise da paisagem: fragmentação e mudanças**. Belo Horizonte: UFMG/Institute of Geosciences, 1998.

SOUZA, E. C. F.; BRANT, A.; RANGEL, C. A.; BARBOSA, E.; CARVALHO, C. E. G.; JORGE, R. S. P.; SUBIRÁ, R. J. Assessing the risk of extinction of Brazilian fauna: a starting point for biodiversity conservation. Diversity and Management 2- 2: 62-75, 2018.

SOUZA, L. B. Leonardo. **How plastic pollution threatens life on Earth**. Autossustentável, 04/06/2018. Available at: <https://autossustentavel.com/2018/06/poluicao-plastico-mares-limpos.html>. Accessed on:2024.

SOUSA, R. **Agrotoxics**. Brazil School. Available at: <https://brasilescola.uol.com.br/geografia/agrotoxicos.htm> . Accessed: 2024.

STUDY MAPS. **Mind maps on climate change**. 10/08/2023. Available at: <https://studymaps.com.br/mudancas-climaticas/>. Accessed on: 2024.

TUNES, P. H. From Tenancy to Metabiosis - Meet 3 organisms that function as ecosystems in the environment in which they live. **Tunes Ambiental**, 20/11/2020. Available at: <https://tunesambiental.com/inquilinismo-3-organismos-que-sao-ecossistemas/#Baleias_-_Ecossistemas_apos_a_morte_-_Metabiose>. Accessed on: 05/04/2023.

TOSSULINO, M. G. P.; MUCHAILH, M. C.; CAMPOS, J. B. The importance of the correct framing of Conservation Units for their effectiveness. *IN*: UNIDADES DE CONSERVAÇÃO - AÇÕES PARA VALORIZAÇÃO DA BIODIVERSIDADE. Curitiba: Environmental Institute of Paraná, 2005.

TYGEL, A. et al. **Agrotoxics Atlas 2023**. Rio de Janeiro, Heinrich Böll Foundation, 2023.

UNEP/CBD/SBSTTA. Economic Valuation of Biological Diversity. IN: Convention on Biological Diversity, July 9, 1996.

UNEP-WCMC and IUCN. Protected Planet Report 2020. UNEP-WCMC and IUCN: Cambridge UK; Gland, Switzerland, 2021.

VILELA, D. A. R.; BARRETO, C.; OLIVEIRA, D. M P. Main threats and measures to safeguard wild animals. IN: CONTROVERSIAL ASPECTS OF CRIMES AGAINST FAUNA. MPMG Jurídico, Fauna Defense Edition, 2016.

VALLADARES-PÁDUA, C. B.; MARTINS, C. S.; RUDRAN, R. Integrated management of endangered species. IN: METHODS OF STUDY IN BIOLOGY OF THE CONSERVATION AND MANAGEMENT OF WILDLIFE, 2 ed. Curitiba: UFPR, 2006.

WARREN, R.; PRICE, J.; VANDERWAL, J.; CORNELIUS, S.; SOHL, H. **Wildlife in an increasingly hot world**. WWF/CLIMATE & ENERGY, 2018. Available at: <https://wwfbrnew.awsassets.panda.org/downloads/wwf_climatespecies_report_portugues_v4.pdf>. Accessed on: 2024.

ZONNEVELD, I.S. **Land evaluation and land(scape)science**: ITC Textbook of photointerpretation. Enschede: ITC, v.7, n.106, 1972.

13 Appendix A
Glossary

The concepts used in this work are based firstly on Brazilian legislation and secondly on specialized literature.

> Abundance of a species - the number of individuals of a species occurring in a given area at a given time.
> Anthropic - resulting from human action.
> Degraded area - where the environment has lost its natural capacity to create benefits for man, vegetation and animals; degradation can be caused by nature or man.
> Watershed - a group of lands where water is collected for a main river and its tributaries; in a watershed, water flows from springs and drains to lower points, forming streams, creeks and brooks that create the main river.
> Biodiversity - the relationship between the number of species and the number of individuals of each species occurring in a given area at a given time.
> Biota - all the flora and fauna of a region.
> Community - groups of populations of different species living in one place at a given time.
> Nature conservation - the management of human use of nature, comprising the preservation, maintenance, sustainable use, restoration and recovery of the natural environment, so that it can produce the greatest benefit, on a sustainable basis, for current generations, while maintaining its potential to satisfy the needs and aspirations of future generations, and ensuring the survival of living beings in general.

- In situ conservation - the conservation of ecosystems and natural habitats and the maintenance and recovery of viable populations of species in their natural environments and, in the case of domesticated or cultivated species, in the environments where they have developed their characteristic properties.
- Chorology - the study of the geographical distribution of living things.
- Ecological corridors - portions of natural or semi-natural ecosystems, linking conservation units, which enable the flow of genes and the movement of biota between them, facilitating the dispersal of species and the recolonization of degraded areas, as well as the maintenance of populations that require areas larger than those of the individual units for their survival.
- Sustainable development - a development model in which the important thing is to generate wealth, distribute it fairly and protect the environment, so that future generations can use natural resources in the same way as they do today.
- Biological diversity - the variability of living organisms from all sources, comprising, among others, terrestrial, marine and other aquatic ecosystems and the ecological complexes of which they are part; also comprising diversity within species, between species and of ecosystems.
- Genetic diversity - genetic variation within a species, being the product of its evolutionary history.
- Ecosystem - system made up of living beings and the place where they live, in perfect balance. example - plants take nutrients from the soil and energy from sunlight; there are animals that feed on plants; these animals serve as food for other animals; when they die, living beings decompose and provide nutrients to the soil, which will again be used by plants - in a life cycle in which each being is of fundamental importance.

- ➢ Environmental education - a set of educational actions aimed at raising individual and collective awareness of the importance of the environment; when people are aware, they change their habits and practice actions that help preserve nature.
- ➢ Endemic - native to a specific region.
- ➢ Threatened species: those whose populations and/or habitats are rapidly disappearing, so as to put them at risk of becoming extinct.
- ➢ Endangered species - a species that is in danger of disappearing completely from the face of the Earth, forever.
- ➢ Exotic species - found outside its natural distribution area.
- ➢ Invasive alien species - which threaten ecosystems, habitats or species; due to their competitive advantages and favored by the absence of predators and the degradation of natural environments, they dominate the niches occupied by native species, especially in fragile and degraded environments.
- ➢ Statistics - characteristics of a sample of a population or community; they are estimators of population parameters.
- ➢ Starvation - a state of lethargy induced in animals by excessive dry heat.
- ➢ Ethology - the study of an animal's behavior in relation to its peers and its adaptation to the environment.
- ➢ Extractivism - a system of exploitation based on the sustainable collection and extraction of renewable natural resources.
- ➢ Fauna - all the species of animals that inhabit a given region.
- ➢ Flora - the set of plant species that make up the vegetation of a given region.
- ➢ Habitat - a favorable environment for the development, survival and reproduction of certain species of animals and/or plants. example - the ecosystem, or part of it, in which a particular living being lives is its habitat.

➢ Biodiversity hotspots - regions of great biological wealth that are extremely threatened.

➢ Irreversible - something that can no longer go back to the previous situation.

➢ Management - any procedure aimed at ensuring the conservation of biological diversity and ecosystems.

➢ Migration - the movement of individuals and/or species, or groups of individuals and/or species from one region to another within the same country; it can be regular or periodic and coincide with the seasons.

➢ Monitoring - is the constant collection of data to track changes and trends in populations and communities.

➢ Parameters - characteristics of the whole population or community.

➢ Management plan - a technical document which, based on the general objectives of a conservation unit, establishes its zoning and the rules that should govern the use of the area and the management of natural resources, including the implementation of the physical structures necessary for the management of the unit.

➢ Population - all the individuals of a species living in a place at a given time.

➢ Preservation - a set of methods, procedures and policies aimed at the long-term protection of species, habitats and ecosystems, as well as the maintenance of ecological processes, preventing the simplification of natural systems.

➢ Integral protection - maintenance of ecosystems free from alterations caused by human interference, with only indirect use of their natural attributes allowed.

➢ Recovery - restoring a degraded ecosystem or wild population to an undegraded condition, which may be different from its original condition.

- ➤ Environmental resource - the atmosphere, inland, surface and underground waters, estuaries, territorial sea, soil, subsoil, elements of the biosphere, fauna and flora.
- ➤ Extractive reserve - an area used by traditional populations who survive on extractivism, subsistence farming and small animal husbandry; its basic objectives are to protect the livelihoods and culture of these populations, and to ensure the sustainable use of the area's natural resources, without private areas.
- ➤ Restoration - restoring a degraded ecosystem or wild population as close as possible to its original condition.
- ➤ Seasonality - variation in the presence of animals due to the seasons. Animals move constantly; the presence of a species in a given area can be seasonal and different from year to year; the results of a survey in one period will certainly be different at any other time.
- ➤ Sustainability - a condition related to the continuity of the economic, social, cultural and environmental aspects of human society; a means of configuring human civilization and activity in such a way that society, its members and its economies can fulfill their needs and express their greatest potential in the present while at the same time maintaining biodiversity and natural ecosystems indefinitely; encompasses various levels of organization, from the local neighborhood to the entire planet. example - for a human enterprise to be sustainable, it has to meet four basic requirements - it has to be ecologically sound, economically viable, socially just and culturally acceptable.
- ➤ Conservation Unit - a territorial space and its environmental resources, including jurisdictional waters, with relevant natural characteristics, legally instituted by the Government, with conservation objectives and defined limits, under a special administration regime, to which adequate protection guarantees are applied.

- ➢ Direct use - that which involves the collection and use, commercial or otherwise, of natural resources.
- ➢ Indirect use - that which does not involve consumption, collection, damage or destruction of natural resources.
- ➢ Sustainable use - exploiting the environment in such a way as to guarantee the continuity of renewable environmental resources and ecological processes, maintaining biodiversity and other ecological attributes, in a socially just and economically viable manner.
- ➢ Buffer zone - the surroundings of a conservation unit, where human activities are subject to specific rules and restrictions, with the aim of minimizing negative impacts on the unit.
- ➢ Zoning - the definition of sectors or zones in a conservation unit with specific management objectives and rules, with the aim of providing the means and conditions for all the unit's objectives to be achieved in a harmonious and effective manner.

14 Annex 1
Law No. 5.197/67 - protection of fauna

Law No. 5.197, of January 3, 1967

Provides for the protection of fauna and other measures.

THE PRESIDENT OF THE REPUBLIC I make known that the National Congress decrees and I sanction the following Law:

Art. 1. Animals of any species, at any stage of their development and living naturally outside captivity, constituting wild fauna, as well as their nests, shelters and natural breeding grounds are the property of the State, and their use, persecution, destruction, hunting or gathering is prohibited.

§ Paragraph 1 If regional peculiarities allow hunting, permission shall be established in a regulatory act of the Federal Government.

§ Paragraph 2: The use, pursuit, hunting or harvesting of species of wild fauna on privately owned land, even when permitted under the terms of the previous paragraph, may also be prohibited by the respective owners, who assume responsibility for overseeing their domains. In these areas, hunting requires the express or tacit consent of the owners, under the terms of articles 594, 595, 596, 597 and 598 of the Civil Code.

Art. 2 Professional hunting is prohibited.

Article 3. Trade in specimens of wild fauna and in products and objects that imply their hunting, persecution, destruction or capture is prohibited.

§ Paragraph 1: Specimens from legalized sources are excepted.

§ Paragraph 2. The collection of eggs, offspring and chicks destined for the aforementioned establishments, as well as the destruction of wild animals considered harmful to agriculture or public health, shall be permitted with a license from the competent authority.

§ Paragraph 3. The mere failure to provide proof of origin of skins or other wild animal products in shipments by land, river, sea or air, which start in or pass through the country, will immediately characterize non-compliance with the provisions of the heading of this article.   (Included by Law no. 9.111, of 10.10.1995)

Art. 4 No species may be introduced into the country without a favorable official technical opinion and a license issued in accordance with the law.

Art. 5. Repealed by Law no. 9.985, of 18.7.2000)

Art. 6 The Government will encourage:

a) the formation and operation of amateur hunting and shooting clubs and societies with the aim of achieving a spirit of association for the practice of this sport.

b) the construction of breeding grounds for the breeding of wild animals for economic and industrial purposes.

Art. 7 The use, pursuit, destruction, hunting or gathering of specimens of wild fauna, when consented to under this Law, shall be considered acts of hunting.

Art. 8 The competent federal public body shall, within 120 days, publish and update it annually:

a) a list of the species whose use, pursuit, hunting or gathering will be permitted, indicating and delimiting the respective areas;

b) the time and number of days in which the above act will be permitted;

c) the daily quota of specimens that may be used, hunted, hunted or caught.

Sole paragraph. Domestic animals that become wild or feral as a result of abandonment may also be used, hunted, chased or caught.

Art. 9 - Subject to the provisions of Article 8 and compliance with legal requirements, specimens of wild fauna may be captured and kept in captivity.

Art. 10: The use, pursuit, destruction, hunting or gathering of wild fauna specimens is prohibited.

a) with mistletoe, slingshots, billy clubs, poison, fire or traps that mistreat game;

b) with shotguns, within three kilometers of any public road or highway;

c) with 22 caliber weapons for animals larger than tapiti (sylvilagus brasiliensis);

d) with traps, consisting of firearms;

e) in urban areas, suburbs, villages and in hydro-mineral and climatic resorts;

f) in official establishments and dams in the public domain, as well as on adjacent land, up to a distance of five kilometers;

g) in the strip of five hundred meters on either side of the axis of railroads and public roads;

h) in areas intended for the protection of fauna, flora and natural beauty;

i) in zoos, parks and public gardens;

j) outside the hunting permit period, even on private property;

l) at night, except in special cases and in the case of pests;

m) from inside vehicles of any kind.

Art. 11 Amateur hunting and shooting clubs or societies may be organized separately or jointly with fishing clubs, and will only function validly once they have obtained legal personality in accordance with civil law and have been registered with the competent federal public body.

Art. 12 - The entities referred to in the previous article must apply for a special license for their members to carry hunting and sporting weapons, for use at their headquarters during the closed season and within the determined perimeter.

Art. 13: In order to hunt, an annual, specific, regional license issued by the competent authority is mandatory.

Sole paragraph. The license to hunt with firearms must be accompanied by a firearms permit issued by the Civil Police.

Art. 14: Scientists belonging to official or unofficial scientific institutions, or indicated by them, may be granted a special permit to collect material for scientific purposes at any time.

§ Paragraph 1 - In the case of foreign scientists, duly accredited by their country of origin, the license request must be approved and forwarded to the competent federal public body, via the country's official scientific institution.

§ Paragraph 2 The institutions referred to in this article, for the purpose of renewing their licenses annually, shall inform the competent federal public body of the activities of the scientists licensed in the previous year.

§ Paragraph 3 The licenses referred to in this article may not be used for commercial or sporting purposes.

§ Paragraph 4. Scientists from national institutions that are required by law to collect zoological material for scientific purposes will be granted permanent licenses.

Art. 15: The Council for the Supervision of Artistic and Scientific Expeditions in Brazil will hear from the competent federal public body whenever there is a matter relating to fauna in the cases being judged.

Art. 16 - A register of individuals or legal entities dealing in wild animals and their products is hereby established.

Art. 17: Individuals or legal entities referred to in the previous article are obliged to submit a declaration of stocks and values, whenever required by the competent authority.

Sole paragraph. Failure to comply with the provisions of this article, in addition to the penalties provided for in this law, shall result in the registration being canceled.

Art. 18: Raw hides and skins of amphibians and reptiles may not be exported abroad.

Art. 19 - The transportation of wild animals, lepidoptera, other insects and their products interstate and abroad depends on a transit permit provided by the competent authority.

Sole paragraph. Material consigned to Official Scientific Institutions is exempt from this requirement.

Art. 20: Hunting licenses will be granted upon payment of an annual fee equivalent to one tenth of the monthly minimum wage.

Sole paragraph. Tourists will pay a fee equivalent to one minimum monthly wage, and the license will be valid for 30 days.

Art. 21: The registration of natural or legal persons referred to in art. 16 shall be made upon payment of a fee equivalent to half a monthly minimum wage.

Sole paragraph. The natural or legal persons referred to in this article shall pay an annual license fee for the various forms of commerce up to the limit of one minimum monthly wage.

Art. 22: Registration of amateur clubs or societies, as referred to in art. 11, will be granted upon payment of a fee equivalent to half a monthly minimum wage.

Sole paragraph. Transit licenses with hunting and sporting weapons, referred to in art. 12, will be subject to the payment of an annual fee equivalent to one twentieth of the monthly minimum wage.

Art. 23: Breeding sites shall be registered for a fee equivalent to two-tenths of the monthly minimum wage.

Art. 24: Payment for the licenses, registrations and fees provided for in this Law will be paid to Banco do Brasil S. A in a special account, to the credit of the Federal Agricultural Fund, under the heading "Fauna Resources".

Art. 25: The Union shall directly supervise the application of the rules of this Law through the specific executive body of the Ministry of Agriculture, or in agreement with the States and Municipalities, and may create the necessary services for this purpose.

Sole paragraph. Supervision of hunting by specialized bodies does not exclude action by the police or the Armed Forces on their own initiative.

Art. 26: All officials in the exercise of hunting control are equivalent to public security agents and are guaranteed the right to bear arms.

Art. 27 - Violation of the provisions of arts. 2, 3, 17 and 18 of this law is a crime punishable by imprisonment from 2 (two) to 5 (five) years. (Edited by Law no. 7.653, of February 12, 1988)

§ Paragraph 1. Violation of the provisions of article 1 and its paragraphs 4, 8 and its subparagraphs a, b, and c, 10 and its subparagraphs a, b, c, d, e, f, g, h, i, j, l, and m, and 14 and its paragraph 3 of this law shall be considered a crime punishable by imprisonment from 1 (one) to 3 (three) years. (Included by Law no. 7.653, of February 12, 1988)

§ Paragraph 2. Any person who, through the direct or indirect use of pesticides or any other chemical substance, causes the perishing of specimens of the ichthyological fauna existing in Brazilian rivers, lakes, reservoirs, lagoons, bays or territorial seas shall incur the penalty provided for in the caput of this article. (Included by Law no. 7.653, of 12.2.1988)

§ Paragraph 3 shall apply to anyone who practices predatory fishing, using a prohibited instrument, explosive, herb or chemical substance of any kind. (Included by Law no. 7.653, of 12.2.1988)

§ 4 (Repealed by Law no. 7.679, of 23.11.1988)

§ Paragraph 5. Anyone who, in any way, contributes to the crimes provided for in the heading and paragraph 1 of this article shall incur the penalties imposed on them. (Included by Law no. 7.653, of 12.2.1988)

§ Paragraph 6 - If the perpetrator of the offense considered a crime under this law is a foreigner, he will be expelled from the country after serving the sentence imposed on him. (Vetoed) The judicial or administrative authority must send a copy of the decision imposing the sentence to the Ministry of Justice within 30 (thirty) days of the final and unappealable decision. (Included by Law no. 7.653, of 12.2.1988)

Art. 28: In addition to the misdemeanors established in the preceding article, the provisions on misdemeanors and crimes established in the Penal Code and other laws, with the penalties contained therein, remain in force.

Art. 29: The following are circumstances that aggravate the penalty, those contained in the Penal Code and the Criminal Contraventions Law:

a) committing the infraction during a closed hunting season or at night;

b) using fraud or abuse of trust;

c) taking undue advantage of a license from an authority;

d) the offense concerns wild animals and their products from areas where hunting is prohibited.

Art. 30: Penalties will be imposed on the authors, whether they are:

a) direct;

b) tenants, partners, squatters, managers, administrators, directors, promisors, purchasers or owners of the areas, provided that it is carried out by agents or subordinates and in the interests of the bidders or hierarchical superiors;

c) authorities who by action or omission consent to the commission of the illegal act, or who commit abuses of power.

Sole paragraph. In the event of simultaneous criminal proceedings for the same fact, initiated by several authorities. The judge shall bring the proceedings together in the jurisdiction in which jurisdiction is established.

Art. 31: Criminal prosecution is independent of a complaint, even in the case of injury to private property, when the property affected is wild animals and their products, working tools, documents and acts related to the protection of the fauna regulated in this Law.

Art. 32: The authorities responsible for initiating, presiding over and carrying out police inquiries, drawing up notices of arrest in flagrante and bringing criminal proceedings, in cases of crimes or misdemeanors provided for in this Law or in other laws whose object is wild animals, their products, instruments and documents related to them, are those indicated in the Code of Criminal Procedure.

Art. 33: The authority shall seize the products of hunting and/or fishing as well as the instruments used in the infraction, and if these, due to their nature or volume, cannot accompany the investigation, they shall be handed over to the local public depositary, if there is one, and failing that, to the one appointed by the judge. (Edited by Law no. 7.653, of 12.2.1988)

Sole paragraph. In the case of perishable products, they may be donated to the nearest scientific institutions, penal institutions, hospitals and/or charitable homes. (Edited by Law no. 7.653, of 12.2.1988)

Art. 34: The crimes provided for in this law are non-bailable and will be investigated through summary proceedings, with the rules of Title II, Chapter V of the Code of Criminal Procedure applying where applicable. (Edited by Law no. 7.653, of February 12, 1988)

Art. 35: Within two years of the enactment of this Law, no authority may allow the adoption of school reading books that do not contain texts on the protection of fauna, approved by the Federal Education Council.

§ Paragraph 1: Primary and secondary education programs must include at least two lessons a year on the subject referred to in this article.

§ Paragraph 2: Radio and television programs must also include texts and provisions approved by the competent federal public body, within the minimum limit of five minutes per week, distributed or not on different days.

Art. 36: The National Fauna Protection Council is hereby established, with headquarters in Brasilia, as an advisory and normative body for the country's fauna protection policy.

Sole paragraph. The Council, directly subordinate to the Ministry of Agriculture, will have its composition and attributions established by decree of the Executive Power.

Art. 37: The Executive Branch shall regulate this Law to the extent deemed necessary for its execution.

Art. 38: This Law shall enter into force on the date of its publication, revoking Decree-Law No. 5894 of October 20, 1943, and any other provisions to the contrary.

Brasília, January 3, 1967, 146th of Independence and 70th of the Republic.

H. CASTELLO BRANCO

Severo Fagundes Gomes

This text does not replace the one published in the DOU of 5.1.1967

*

Printed by Books on Demand GmbH, Norderstedt / Germany